T0136468

Forestry and the
Forest Industry in Japan

Edited by Yoshiya Iwai

Forestry and the
Forest Industry in Japan

UBCPress · Vancouver · Toronto

09 08 07 06 05 04 03 02   5 4 3 2 1

Printed in Canada on acid-free paper ∞

National Library of Canada Cataloguing in Publication Data

Main entry under title:

Forestry and the forest industry in Japan

   Includes bibliographical references and index.
   ISBN 0-7748-0882-9 (bound); ISBN 0-7748-0883-7

   1. Forests and forestry – Japan. – I. Iwai, Yoshiya, 1945-

SD225.F67 2002   634.9'0952   C2002-910390-8

Canadä

UBC Press gratefully acknowledges the financial support for our publishing program of the Government of Canada through the Book Publishing Industry Development Program (BPIDP), and of the Canada Council for the Arts, and the British Columbia Arts Council.

Set in Stone by Artegraphica Design
Printed and bound in Canada by Friesens
Copy editor: Francis Chow
Proofreader: Janet Dimond
Indexer: Christine Jacobs

UBC Press
The University of British Columbia
2029 West Mall
Vancouver, BC V6T 1Z2
(604) 822-5959 / Fax: (604) 822-6083
www.ubcpress.ca

Contents

Part 2: Forest and Wood Products Industries in Japan

Part 3: New Trends for Forestry in Japan

Figures and Tables

Tables

Exchange rates for Japanese yen in US dollars, selected years, 1960-99.

Year	US \$/¥1,000	Year	US \$/¥1,000
1960	2.79	1984	4.10
1965	2.77	1986	6.26
1970	2.80	1988	7.80
1972	3.31	1990	7.08
1974	3.32	1992	8.01
1976	3.42	1994	10.06
1978	5.14	1996	8.88
1980	4.93	1998	7.81
1982	4.26	1999	8.97

Source: Naikakufu (1985, 2001), Keizai-yoran.

Preface

Sixty-seven percent of Japanese land is covered with forest, 40% of which is the result of afforestation. Despite this wealth and a history of silviculture that goes back hundreds of years, Japanese forestry has lost international competitiveness and faces a host of problems. Self-sufficiency is low, and 80% of the Japanese demand for timber is filled by imports. While the domestic cutting volume of forest has decreased rapidly, the annual growing stock has grown to twice the annual cutting volume. Forest areas left unplanted after harvesting have also been increasing, with serious implications for erosion control. Other countries have similar problems of forest management, but the situation in Japan appears to be dramatically worse compared with some of them.

In recent years, researchers in the areas of forest policy, forest economics, and forest management have gathered to exchange information and research results at international conferences and through bodies such as the International Union of Forestry Research Organizations (IUFRO). It appears, however, that information on forestry in Japan has not been available to many people in other countries.

Planning for this volume began three years ago, with the goal of disseminating information on Japanese forestry. From the many Japanese forestry researchers, 15 individuals working in the western part of the country contributed to this publication. They describe the real state of forestry and the forest industry in Japan and focus on its problems. The authors hope that this book will be read by researchers and others interested in forestry around the world, and that it will contribute to finding solutions to the problems of forestry and forest management in different countries, including Japan.

Yoshiya Iwai
Kyoto
August 2001

Acknowledgments

I would like to thank Tsuguo Mitsui, former chairman of the Japan Wood-Products Information and Research Center (JAWIRC) in Tokyo, and Hikojiro Katsuhisa, manager of the Seattle branch of JAWIRC, through whom this project came to the attention of UBC Press. Mr. Katsuhisa introduced me to Randy Schmidt, an editor at UBC Press, who together with four reviewers read the manuscript intensively and considered it worthy of publication. I would also like to thank Holly Keller-Brohman, the managing editor at UBC Press, for her work on the book.

The cooperation of these individuals and their efforts to get this book published are greatly appreciated.

Introduction

Yoshiya Iwai

This book consists of three parts.

Part 1 looks at the development and characteristics of forestry in Japan, where 58% of forests are privately owned. Historically, tree plantations started with such privately owned forests, which for a long time played an important role in supplying wood for the domestic market. Chapter 1 describes forest ownership in the feudal era, and the formation and establishment of private forests in the modern age.

As the Japanese generally love high-quality wood without knots – in other words, high-priced wood – private forest owners have undertaken very intensive silviculture to raise such trees, as described in Chapter 2. Since 1990, however, the demand for high-quality wood has been decreasing because of economic recession, and intensive silviculture has become unprofitable. Chapter 3 focuses on the factors contributing to unprofitability in privately owned and managed forests from the standpoint of silviculture costs and internal rates of return.

The private forest owners established a forest owners' association in 1907. Since 1970, more and more small forest owners have been commuting to work in cities and have been unable to work in their own forests. Chapter 4 describes the role played by the forest owners' association in caring for forests on behalf of the owners. Chapter 5 focuses on the workers employed by big private forest owners and the owners' association to take care of the forests. Such workers have been declining in number in recent years.

Whereas Chapters 1 to 5 deal with private forestry, the next three chapters discuss forestry and the forest industry from a national perspective. The national forest was established in 1899 and has supplied much wood to the market. Since 1970, however, it has become unprofitable. Chapter 6 describes the national forest management system and the changes it is undergoing.

Every five years, the national government prepares nationwide forest plans that cover the national forest and private forests for a 10-year period. The plan sets out goals for harvesting volume, planting area, and so on, but results fall short every year because of the decline in forest production. The national forest plan and its problems are discussed in Chapter 7.

Chapter 8 describes the forest policies of the national government and local governments. Such policies are very important because private forest owners are strongly supported by government subsidies for forest road construction and planting.

Part 2 examines the forest industry and the market for forest products in Japan. Logs harvested from domestic forests are gathered in log markets, where they are classified by size and quality and sold to sawmills by auction. Chapter 9 discusses the characteristics of this distribution system, while Chapter 10 discusses the characteristics of the sawmill industry. Sawmills in Japan, especially those processing domestic logs, are much smaller than those in other countries from which Japan imports sawnwood. Furthermore, these mills do not have facilities for grading or kiln-drying, and have no competitive power in the market.

The two industries that consume the most wood materials in Japan are home building and pulp and paper. Chapter 11 discusses the structure and behaviour of the home-building industry, while Chapter 12 describes the pulp and paper industry, including the system for purchasing raw materials.

Large enterprises, such as the pulp and paper industry, depend heavily on raw materials from foreign countries. Chapter 14 focuses on wood trading and discusses the reasons why imported wood has a large share of the Japanese market. Although domestic wood cannot compete in the large industrial market, more domestic wood than imported wood is used in some local markets, such as housing materials. Chapter 13 describes the local markets from the perspective of domestic wood supply and demand.

Part 3 is about emerging trends in forestry in Japan. Mountain villages that depend on forestry face a crisis because of the depressed state of the industry. Chapter 15 describes how such areas can be revitalized by new, non-forest-related businesses. Finally, Chapters 16 and 17 discuss the new relationship between people and forests resulting from the growing interest in forests, forestry, and wild animals on the part of city dwellers.

To deepen the reader's understanding of these topics, the following sections provide more information about the historical flow and changes in forestry and the forest industry in Japan.

Small Forest Ownership

Fifty-eight percent of Japan's forests is privately owned, 31% is national forest, and the rest is owned by local governments. A part of the privately owned forest is owned by paper mills and mining companies (company forest), but harvesting is hardly done today. An overwhelming part of the privately owned forest is owned by small-scale farmers. Some privately owned forest also existed during the Edo era (1603-1867), also known as the Tokugawa period, and many of the communal forests were owned by farmers' groups, who obtained from them grass as agricultural fertilizer, as well as firewood.

The land revolution during the Meiji period (1868-1912) resulted in part of the communal forest being divided among individuals, so that each farmer owned a small portion of the forest. The rest of the communal forest was transferred to national forest. Even today privately owned forests in Japan exist on a small scale. The problems of small-forest management in Japan began, therefore, during the Meiji period. When agrarian reform was initiated in 1945, the large landownership system of cultivation fields was dissolved and many tenant farmers became owner farmers, but the area of individual cultivation land became smaller. Until 1950, agricultural mountain villages had a large population, and most of the farmers made their living by rice cultivation, cattle breeding, fuelwood production, and some industrial wood production.

Increase in Demand for Wood

Japan was self-sufficient in wood before 1920 and during the Second World War, and wood was imported in the interwar period between 1920 and 1940. Wood supply was stable until about 1950, when during the Korean War Japan's economy recovered from the damage of the Second World War and the demand for wood began to increase. In the 1960s a period of high economic growth began, and the demand increased even more.

The forest planning system began in 1951 in response to the increased demand for wood, in order to develop the forest resource promptly and to stabilize and expand domestic supplies. With increased wood production, sawmills became specialized and log distribution was rationalized. Log markets were set up to provide a system for efficient distribution of logs from forests to sawmills. Sawmill companies stopped logging and began concentrating on sawmill operations. The size of sawmills was relatively small. The increased demand for wood triggered a rise in the price of wood; both forestry activity and the profit from forestry also increased.

The Forestry Basic Law was established in 1964 and various policies led to an increase in gross forest production. An enormous investment was made in building forestry roads in order to improve production. The harvest volume of the national forest increased rapidly, and most of the proceeds from this were placed in the national general account, with none of the revenue reserved in the national forest account. Later, an independent accounting system for the national forest was created, but because of declining forest revenue, the national forest carried a large debt.

The Fuel Revolution and the Expansion of Man-Made Forest

In the 1960s, two major changes took place: changes in the use of fuel, called the fuel revolution, and the outflow or migration of people from mountain villages to big cities.

Before the fuel revolution, until about 1960, fuelwood and charcoal were the main fuel for daily life not only in mountain villages but also in many cities in Japan. Broad-leaved trees and pine wood were used for fuel, and the income from selling these was sufficient for mountain villages. As electricity, gas, and oil rapidly became available as fuel, however, the demand for fuelwood decreased rapidly. In mountain villages today, it is becoming rare to see wood used as the primary fuel for heating or cooking, unlike in European countries. Since broad-leaved trees were useless as fuel, they began to be sold as pulpwood, and sugi (Japanese cedar, *Cryptomeria japonica*) and hinoki (Japanese cypress, *Chamaecyparis obtusa*) were planted to replace them.

The price of sugi and hinoki surpassed that of fuelwood and pulpwood because of the increased demand for wood as building material, and natural forests of broad-leaved trees rapidly changed to plantation forests of sugi and hinoki. Tree planting reached its peak in 1961, although it continued at a high rate for 10 more years. During this period, sugi and hinoki forests promoted weed overgrowth as they consisted of young trees (one to eight years old), and provided feeding places on a large scale for wild animals, especially deer. Thus the population of wild animals increased, resulting in tree damage through browsing in recent years.

Underpopulated Mountain Villages

During the 1960s the outflow of people from mountain villages to big cities caused a serious depopulation phenomenon. As Japan's economy began to catch up with those of the developed Western countries, manufacturing industries in big cities needed a growing workforce, which was supplied by the agricultural mountain villages. The outflow of people to the big cities was desirable because these villages had a large population and a surplus workforce for forestry and agriculture due to the decreased demand for fuelwood and the introduction of chemical fertilizer and mechanized agriculture. The rapid outflow, however, became a serious problem and resulted in underpopulated villages.

Wood Imports

As mentioned earlier, although harvesting was actively being carried out, the price of wood continued to rise. Until 1961, priority in the use of foreign currency was given to importation of iron ore and oil, and there were strict limitations on wood imports. The rapid increase in the price of wood affected the nation's daily life, however, so the government liberalized wood importation and gradually abolished import duties. Free importation of wood actually began 30 years earlier than the liberalization of rice importation in Japan, which started in 1998.

After liberalizing wood importation, the government improved harbour facilities and developed large sawmill complexes around the harbours. Logs were imported from overseas and systems for mass production and manufacturing by large-scale sawmills were put in place. (The sawmills, however, were much smaller than those in North America, from where Japan imported much sawnwood later on.)

In 1969 imported wood made up 50% of the total wood consumption in Japan for the first time. Tropical wood from Southeast Asia accounted for most of the logs and sawnwood imported in the 20 years from 1961, followed by wood from North America. From the mid-1980s, however, North American wood took first place. After 1985, most of the tropical wood imports were lauan that had already been processed into plywood, and most of the wood from North America was sawnwood for housing construction. Pulpwood imports increased a little later than sawnwood, but after 1970 the amount of pulpwood imports increased largely because of ships put into operation exclusively for chip imports.

The use of broad-leaved trees in Japan changed from fuel to pulp; these trees gradually became useless because of the increasing volume of chip imports. Most of the broad-leaved trees that were not replaced by man-made forests of sugi and hinoki were abandoned without management. In recent years, these abandoned forests have increasingly become the focus of work by volunteers from neighbouring cities.

The Advantage of Imported Wood

Why did Japan import a large amount of wood even though it had a rich and active forest production? The first reason was the low price of imported wood; the second was the stable supply of uniform wood. These were important considerations for those accounting for the high demand for wood in Japan – the pulp, sawmill, housing, and plywood industries – as their business scale became larger and consumed more wood.

In contrast, Japan's domestic wood was more expensive because of high silviculture costs, especially for weeding in rainy season. Since privately owned forests were small and the forestland is very steep, logging costs were high. Moreover, forests were owned mostly to maintain a farmer's property, so trees were not always harvested even during the cutting period, and the supply of domestic wood was not stable.

Stagnation in Domestic Forestry

With the increase in imported wood, the stumpage price of domestic wood gradually decreased. This tendency started from the late 1960s, and after 1970 the stumpage price appeared to be low because of the strength of the yen in the foreign exchange market (see the appendix on p. x for a list of historical exchange rates for the yen in US dollars). The depression and fall in stumpage prices with the rising wages of forest workers led to a drop in the profitability of forestry. Forestry activities declined and sawmills were forced to adjust to the situation. The harvest volume in Japan reached its peak in 1966 and has decreased steadily since. A great gap between planned and actual production was inevitable. The difficult situation did not uniformly affect forest production activities all over the country, however; several examples of this are mentioned later under "Survival Strategy."

It should be noted that only the price of domestic wood that was not of high quality declined. An explanation of this can be found in Japanese taste and culture. The Japanese language uses three different kinds of letter groups, one of which, the kanji, originated from China. Each kanji character has a specific meaning by itself. About 10,000 kanji characters are used today, and those related to trees and wood are the most common. From ancient times, the basic structure of Japanese buildings such as shrines and temples, and even houses, was all made of wood. Even today, 40% of Japanese houses are made of wood. In other words, the Japanese made their living by using wood and a culture relying greatly on wood was formed. Wood is therefore very special to the Japanese taste.

Japanese are fond of wood that is knotless, lustrous, and fine-grained for houses and furniture. Although it is certain that Europeans and Americans also like knotless wood better than wood with knots, their taste cannot be compared with that of the Japanese, whose regard for knotless, lustrous, and fine-grained wood – that is, high-quality wood – is the greatest in the world. There were many kinds of imported wood similar to sugi but not much imported wood similar to hinoki, which is especially lustrous. Thus the high-quality wood of hinoki (as well as that of sugi, of course) became highly regarded.

Survival Strategy

When stumpage prices began to decline with the rising wages of forest workers, various survival strategies were adopted in forestry and the forest industry. The first strategy was to boost the production of high-quality wood, mostly hinoki, through silvicultural practices. Dense planting, intensive pruning, and regular thinning were done to produce high-quality wood in some areas.

A survival strategy was also implemented in the sawmill industry. For example, in areas that produced scaffolding boards, small sawmills started a joint order system in cooperation with one another. Another example involves sawmills selling processed building materials to local carpenters and small housing companies. Houses built in the Japanese countryside use various sizes of housing materials that differ from those used for houses in urban areas. For such houses, it is not suitable to use simple-sized wood materials processed by large sawmills such as harbour sawmills that process imported logs. Small sawmills that can process small orders can respond better to the demand for housing materials of various sizes. This survival strategy was adopted by sawmills dealing in domestic wood.

The demand for local wood is small, however, and it is not certain that many of the sawmills processing domestic logs and depending on it can survive. Sawmills that failed to adopt any survival strategy could not sustain their activities and were forced to close. Because many sawmills processing domestic logs went out of business, the demand for domestic logs decreased.

Retrogression in Forest Management
Because of the unfavourable situation in forest production, owners lost interest in forest management. As mentioned earlier, the area of planted sugi and hinoki forest was enlarged beginning from the mid-1950s to the period of high economic growth, and instances were noted where the forest had not been thinned although the trees were old enough for thinning.

In the early 1980s, the national government and local governments implemented several policies to promote thinning. Forest owners' associations recommended thinning to forest owners, and some of them developed new ways of utilizing thinned logs, such as producing materials for log houses. Despite these efforts, not much thinning was carried out. The forests of sugi and hinoki that had been abandoned remained unthinned, resulting in damage to the understory of vegetation; in addition, forest topsoil was washed away into rivers or the forest itself became easily damaged by wind or snow. Such lack of thinning finally led to volunteer activities to take care of the forests.

In 1991 the national government initiated a new policy based on river basins to revitalize forestry. This comprehensive system centred on individual rivers, and logs collected in the forest of the upper region were sent down the river to be processed and consumed. A large investment was made in this system.

On the other hand, the demand for high-quality domestic wood has decreased significantly as a result of the recent economic depression. Forestry intended for production of high-quality wood faces a difficult situation. The fact that laminated wood processed from imported European spruce has gained the largest share of the high-quality wood market indicates that the definition of high-quality wood has been changing. Now "high quality" refers to materials made with highly developed technology that conform to strict dimensions and standards for stability as building materials. Under these conditions, forestry and the domestic wood market have entered an even more severe phase.

Forests and the Environment

The development of the heavy machinery and chemical industries, and the concentration of population in cities during the period of high economic growth, resulted in heavy pollution of industrial areas and big cities. Drainage, smoke from factories, and exhaust from cars frightened city dwellers, and a campaign against pollution became active not long before the notion of environmental safeguards arose. The pollution problem soon revived interest in forests. The global problem of decreasing forest resources, mainly tropical rain forest, and global warming due to an increase in carbon emissions, drew the interest of city dwellers to forests even more. In 1972 the national forest's management policy changed: priority was shifted from wood production and greater importance was attached to the public benefit of forests, resulting in widespread change in how forests were cared for.

The increasing number of people who live in cities and who are interested in hiking, camping, and mountain climbing shows that the time for improving the recreational function of forests has arrived. Some agricultural and mountain villages, which could not earn a living just by agriculture and forestry, began to take up forest recreation as a village revitalization and city interchange program, which made progress through the 1980s.

In the history of Japan, it has been extremely rare that city residents have been directly concerned with forestry; in fact, the historically weak relationship between city residents and forests was one of Japan's features. The new developments through the 1980s appear to be creating a new relationship between forests and city dwellers. In 1984 a profit-sharing system for the national forest was introduced, in which residents of cities invested money. In the 1980s, a campaign was initiated for volunteers to take care of forests that were not well cared for. In the early 1990s, young businessmen from cities flocked to the recruitment of forest workers by the forest owners' association, which was suffering from an inadequate workforce. The 1990s were also the period when rental forests similar to *Klein Garten* (a kind of small rental plot of land for leisured agriculture) appeared, and a system for citizens to enjoy free forestry work was begun.

A new culture in which people in cities are interested in forests or forestry and are getting involved in forestry activities is about to begin, but it is difficult for these people to tend all the forests in Japan and put them under good management. The main problem is that forest management in Japan is not sustainable in general, and most Japanese forests are becoming unhealthy.

Part 1
Forestry and Forest Policy in Japan:
Historical Development and
Current Situation

Distribution of forests in Japan

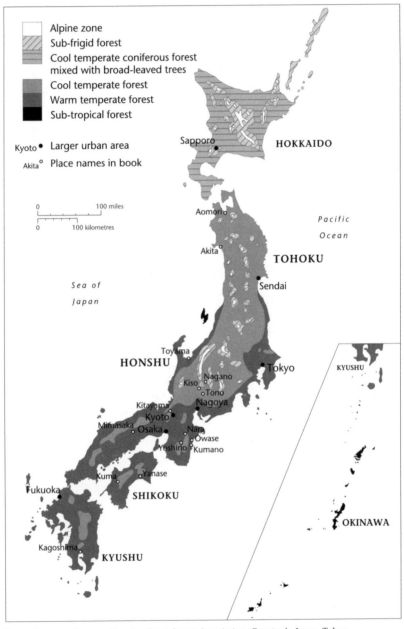

Legend:
- Alpine zone
- Sub-frigid forest
- Cool temperate coniferous forest mixed with broad-leaved trees
- Cool temperate forest
- Warm temperate forest
- Sub-tropical forest

Kyoto ● Larger urban area
Akita ○ Place names in book

0 — 100 miles
0 — 100 kilometres

HOKKAIDO

Sapporo

Pacific Ocean

Aomori

Akita

TOHOKU

Sendai

Sea of Japan

Toyama

HONSHU

Nagano
Kiso
Tono
Nagoya
Tokyo

KYUSHU

Kitayama
Kyoto
Mimasaka
Osaka
Nara
Owase
Yoshino Kumano

Kuma
Yanase

Fukuoka

SHIKOKU

Kagoshima

KYUSHU

OKINAWA

Source: Japanese Overseas Forestry Consultants Association, *Forestry in Japan*, Tokyo.

1
The Development of Japanese Forestry
Junichi Iwamoto

The Japanese have derived benefits from forests throughout history. For example, wood was the primary fuel in Japan until the 1950s. It has also been especially important as a traditional Japanese building material. The world's oldest currently existing wooden building is found in Japan. Built as a temple named Horyuji, it is a complex of large-scale timber-style construction using old-growth hinoki (Japanese cypress, *Chamaecyparis obtusa*) lumber and is over a thousand years old. The oldest buildings in the complex were constructed in the late seventh century.

For a long time, wood was the only feasible building material. Timber construction and wood processing technology were highly developed even in ancient Japan. Besides temples and shrines, ordinary houses were built of wood, which is well suited to Japan's humid summer climate, and wooden houses continue to be popular. Thus, forests have provided generations of Japanese with building material.

By the late 17th century, plantation forestry had begun in Japan; before this, abundant natural forests had provided timber. Timber plantations were started after clear-cutting or slash-and-burn cultivation. From then until now, artificial planting has involved mainly sugi (Japanese cedar, *Cryptomeria japonica*) and hinoki.

Artificial planting was actively practised in three periods of Japanese history during which rapid urbanization and economic expansion occurred. Because the construction of traditional Japanese-style buildings requires huge quantities of timber products, urbanization has been a main factor in the emergence of widely practised artificial planting. The first period began as early as the late 17th century. The second period, characterized primarily by rapid industrial growth, began at the start of the 20th century. The third period was the 1950s, during which the postwar Japanese economy made a quick recovery. Economic growth created high demand for timber, and high timber prices motivated people to plant trees for investment.

Forest Management by Feudal Lords

During the Tokugawa period (1603-1867), the three principal types of forest land tenure were the feudal lord's tenure, communal tenure, and individual tenure. Individual tenure failed to develop because individual land ownership was prohibited in principle by the Tokugawa Shogunate. Consequently, almost all Japanese forest land tenure was either the feudal lord's tenure or communal tenure.

Access to the forests owned by feudal lords was strictly limited, and those who stole timber from the feudal lords' forests were severely punished. A typical example of forest owned and managed by a feudal lord was the Kiso area (Figure 1.1), where the forest was owned and managed by a relative of

Figure 1.1

Two major cities (Edo and Osaka) and three main timber-producing districts (Kiso, Yoshino, and Owase) in 17th-century Japan.

the Tokugawa Shogun. Before the Tokugawa period, Kiso was covered with thick forest. After more than a hundred years of war among feudal lords, peace began at the start of the 17th century. Commerce and industry developed, creating a great need for wood. By the late 17th century, Kiso's forest resource had deteriorated. Its feudal lord therefore carried out the first reform of the forest management system in 1665, instituting seedling protection, strengthening of patrols, and selective cutting. The reform reduced Kiso's timber production by half and cut the feudal lord's income severely. Only a few years later, he ordered an increase in timber production for financial reasons.

Although the first reform failed, the second reform was planned in 1724. In this reform, timber production was reduced by more than 60%. This second reform succeeded, and reduced timber production continued for 30 years, allowing Kiso's forest resource to recover remarkably. It is notable that although forest resource deterioration began as early as the 17th century, forest conservation has also been practised since then. In Kiso, forest inventory and production planning began in 1779.

Forestry on the Common Lands

Japan is a mountainous country where extensive common lands existed during the Tokugawa period. At that time most Japanese made their living by agriculture, managing uncultivated mountainous common lands surrounding their villages. The common lands provided a wide variety of forest products, such as fuelwood or decayed plants for fertilizer.

In the late 17th century, intensive forestry with artificial planting was begun by members (farmers) of the commons in response to increasing demand for wood, and was practised from the 17th to the 18th century. People planted valuable conifers such as sugi or hinoki on the common lands. Each plantation was relatively small because farmers planted seedlings when they were free from agricultural work.

Thus, plantation forestry has a long history in Japan. During the feudal era it was not widely practised but was limited to areas that had good conditions for growing valuable conifers and for timber marketing. One of these, the Yoshino area, is located in the upper district of the Yoshino River, where plantation forestry commenced between the late 17th and early 18th centuries (Figure 1.1). Before then, people historically made a living using the abundant natural forest resources. Timber production from the natural forest increased rapidly in the early 17th century, with Yoshino timber being rafted down the Yoshino River and shipped to timber markets in Osaka, Japan's second largest city.

Because of deterioration of the natural forest resource, artificial planting was done. Although Yoshino was ruled directly by the Tokugawa Shogunate, most of its forests were managed by local villages whose custom approved the private ownership of trees planted by individual villagers. This custom encouraged people in Yoshino to plant trees.

There were two other reasons for the establishment of plantation forestry in Yoshino. First, the timber distribution system, especially for young (20- to 30-year old) timber, had already been established. Second, timber sales were conducted in Yoshino, and timber younger than the cutting age could be sold for cash. Tree planting was done by individual farmers. Frequently, however, farmers sold timber to rich landowners or merchants who held the trees until they were old enough to be sold at a high price. Because these rich landowners and merchants lived outside the Yoshino area, silviculture was entrusted to the local people, who had a highly developed technique.

Another typical area where plantation forestry was carried out on communal forestland was the Owase area, near the seashore, where weather conditions are very suitable for forestry (Figure 1.1). Owase flourished as an important port located on the sea route between Edo and Osaka. It was a prosperous port town, home to many rich local merchants who invested in plantation forestry beginning in the 17th century.

Ruled by a feudal lord, Owase followed Yoshino's custom of approving the private ownership of trees planted by individual villagers. Behind this custom was the feudal lord's desire to promote forestry, raise forestry production, and tax the products. In Owase, artificial planting was widely practised from the early 17th century. Residents produced large quantities of charcoal from broad-leaved trees and planted valuable conifers. Tree planting was initially done by individual farmers. Later on, relatively large timber plantations were started by rich local merchants using local labour.

The Emergence of Plantation Forestry

The most important factor in the emergence of Yoshino and Owase plantation forestry, which began in the 17th century, was the process of urbanization occurring at the time. After the Ohnin War, which started in 1467, there were many battles between feudal lords. Lord Nobunaga Oda reunified Japan in 1568 but committed suicide in 1582, during a rebellion by one of his retainers. Japan was reunified by Ieyasu Tokugawa in 1603, establishing the Tokugawa Shogunate, which continued until 1867, a period referred to as the "Tokugawa period" or Edo era.

The Tokugawa Shogunate divided Japan into approximately 300 territories, each ruled by a feudal lord. The feudal lords built castles around which towns were constructed. The lords let merchants live around their castles, giving them favourable treatment to allow the towns to prosper.

Many cities were constructed throughout the country during the 17th century. Among them, Edo (later Tokyo) and Osaka were outstanding in size, even compared with the world's large cities. In 1801 London's population was 860,000 and that of Paris was 540,000; Edo's was 1,000,000. It is notable that Edo was the world's largest city from the late 17th to the early 19th century.

Edo flourished as the political centre of the Tokugawa Shogunate. Half of its population were warriors (*bushi*), who were virtual bureaucrats, and half were engaged in commerce and industry. Osaka was Japan's most prosperous commercial city. Between Osaka and Edo was a regularly scheduled sea route across which moved large quantities of goods.

The development of these cities during the 17th century was accompanied by rapid growth in the demand for timber, the main material for construction. Since many wooden structures were built close to each other in the cities, there were frequent catastrophic fires, which further increased the demand for timber.

Natural forest resources deteriorated in the 17th century, setting the stage for Japanese plantation forestry to begin. The isolationist policy of the Tokugawa Shogunate also contributed to this development: although commerce within Japan was quite active, foreign trade was prohibited, thereby blocking timber imports during the Tokugawa period.

Forestry in the 20th Century
The Meiji restoration (1850s to 1871), one of Japan's most significant events, ended the Tokugawa Shogunate. Japan began to modernize and Western culture was introduced. One of the main reforms was the establishment of land ownership. Land "ownership" was not clearly distinguished in the feudal era, and people had only a vague awareness of land "tenure."

To stabilize revenue, the new government established the national forests, consisting chiefly of communal and former feudal lords' forests. The government recognized that many communal forests were not being utilized, and merged them with the national forests.

In 1899 the national forests initiated a large project to establish a forest management system. Initially they were divided into "necessary" and "unnecessary" forests. The "unnecessary" forests were sold for funds to establish the forest management system. Next, the "necessary" forests were surveyed and management plans were developed. Tree planting and forest road construction were done. This project lasted for 16 years and established the basis of the forest resources in today's national forests.

During this period, Japan was rapidly changing into an industrialized country and the increasing demand for wood in the expanding economy resulted in destructive cutting nationwide, leading to many disasters. The First Forest Law was therefore passed in 1897, the first modern law in Japan established for the purpose of forest regulation and control. In 1907, however, the Second Forest Law changed forest policy by shifting the emphasis from regulation to utilization, as economic growth during this period required efficient timber production.

The use of communal forests began to change early in the 20th century. Earlier, communal forests were used mainly for gathering fuelwood or manure. As part of the effort to increase national power, however, the government promoted tree planting in these forests from the beginning of the 20th century. Japan's victorious wars with China in 1894-95 and Russia in 1904-5 required substantial sacrifices from the Japanese people, chiefly from

Figure 1.2

Area regenerated by artificial planting on an annual basis, 1950-2000.

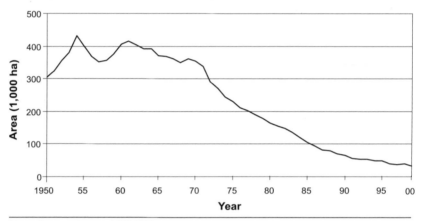

Source: Forestry Agency, 1951-2000.

Figure 1.3

Japan's declining self-sufficiency in wood products, 1955-2000.

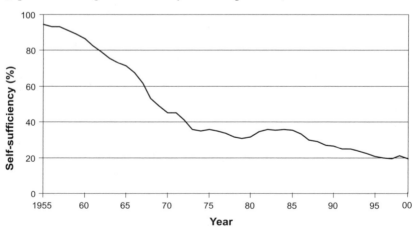

Source: Forestry Agency, 1956-2001.

rural residents because Japan was still basically an agricultural country. To enhance rural economic strength, therefore, the government promoted afforestation of communal lands.

Forest resources were overused during the Second World War, and immediately after Japan's defeat in 1945, the most important problem was food shortage. This crisis was overcome, and beginning in the 1950s, Japan rose as a democratic country. Its economy boomed as a result of US procurement demands during the Korean War (1950-53). During this time, artificial planting was resumed, stimulated by rapidly increasing demand for timber. Thus began the third boom of artificial planting in Japanese history.

The annual increase in artificial plantation area peaked in 1955 and remained high until 1970 (Figure 1.2). During the 1950s and 1960s, artificial planting was done mainly by small-scale family farmers. After the 1960s, forests regenerated by artificial planting began to decrease. In contrast, timber imports began increasing in the 1960s and timber produced in Japan lost its competitive advantage, mainly because of rising wages (Figure 1.3).

Because of many difficulties from the perspective of timber production (low profitability, shortage of forestry workers, etc.), the greater part of plantation forests have been left without proper care and management. On the other hand, there is growing realization of the environmental values of forests (prevention of erosion, clean water supply, preservation of biodiversity, etc.), and these values have been enhanced year by year. Compared with other industrialized countries, however, Japan has been slow to make the transition to environmentally oriented "new forestry."

References

Funakoshi, Shoji (1981). *Nihon no Ringyo, Rinsei*. 341 pp. Nourin Toukei Kyoukai, Tokyo (in Japanese).

Handa, Ryoichi, and Sumiyoshi, Ariki (1990). "Nihon rinsei no tenkai." In Handa, R., ed., *Rinseigaku*, 51-94. Bun-eido Publishing Co., Tokyo (in Japanese).

Nihon Ringyo Gijutsu Kyoukai, ed. (1972). *Ringyo Gijutsu-shi dai-1-kan*, 1-173. Nihon Ringyo Gijutsu Kyoukai, Tokyo (in Japanese).

Tokoro, Mitsuo (1975). *Kinsei Ringyo-shi no Kenkyu*. 858 pp. Yoshikawa-kobunkan, Tokyo (in Japanese).

Totman, Conrad (1989). *The Green Archipelago: Forestry in Preindustrial Japan*. 290 pp. University of California Press, Berkeley and Los Angeles.

2
Silviculture in Japan
Mitsuo Fujiwara

Some say that Japanese silviculture is characterized by plantation forests –
even-aged plantations developed using clear-cutting. This characteristic is
clearly reflected in the fact that two-thirds of Japan's total land area is covered by forests, 40% of which are planted. There is a continuing argument,
however, regarding whether a plantation forest silvicultural system exists
and has been implemented. Some argue that manuals exist for silviculture,
at least for sugi (Japanese cedar, *Cryptomeria japonica*) and hinoki (Japanese
cypress, *Chamaecyparis obtusa*), and that planting and tending according to
these manuals would certainly improve forest productivity. Others claim
that a silvicultural system that channels the potential of the forest's natural
forces has not yet been formulated, and that highly technical silviculture is
impractical today, as it is unclear who is responsible. Another opinion is
that silviculture techniques develop gradually, since silviculture is strongly
influenced and limited by social considerations, the economy, and the allocation of techniques.

Ninety percent of Japan's plantation forests were planted after the Second
World War, and most are still in their first generation. Reviewing Japanese
silviculture, therefore, means tracing the changes in the plantation forests'
silviculture and growth. This chapter focuses on the development of silviculture in these plantation forests, mainly in the postwar period. Silvicultural
techniques that have been introduced, their socio-economic background,
and accompanying problems are discussed.

Major Forestry Areas before the Second World War
Forests have been planted in Japan since the end of the 16th century, and
silvicultural techniques developed during the past four centuries have greatly
influenced modern silviculture. Despite the big differences that exist between national and private forests, even before the Second World War several forestry areas were well known for their silvicultural techniques and
the structure of their forest stands (Figure 2.1). Most of the major national

Figure 2.1

Traditional forestry areas in Japan.

Aomori
(Thujopsis dolabrata)

Akita
(Cryptomeria japonica)

Kiso
(Chamaecyparis obtusa)

Nishikawa

Kitayama

Sanbu

Chizu

Tenryu

Yoshino

Hita

Owase

Kito

Yanase
(Cryptomeria japonica)

Obi

Yakushima
(Cryptomeria japonica)

forests were almost pure natural coniferous forests that had been preserved by the strict cutting restrictions of the feudal Edo era governments and had later been incorporated into the national forests during the Meiji era. Examples include the so-called three most beautiful forests: the forests of

Thujopsis dolabrata var. *hondae* in Aomori, the sugi forest in Akita, and the hinoki forest in Kiso. The sugi forest in Yanase is also famous.

Alternatively, private forests were planted along river basins in traditional forestry areas, including Nishikawa (Saitama Prefecture), Sanbu (Chiba Prefecture), Tenryu (Shizuoka Prefecture), Owase (Mie Prefecture), Yoshino (Nara Prefecture), Kitayama (Kyoto Prefecture), Chizu (Tottori Prefecture), Kito (Tokushima Prefecture), Hita (Oita Prefecture), and Obi (Miyazaki Prefecture). Among these, Yoshino has the longest history of planting, dating back to the end of the 16th century. Although planting gradually developed in other areas later, forestry areas did not expand rapidly until the end of the 19th century.

The prewar characteristics of private, advanced forestry areas were influenced by three factors. First, in forestry areas where logs were transported by rafting, river conservation was necessary to expand the plantations. Much of the forestland was purchased and reforested by downstream merchants, who could afford to invest in both reforestation and river conservation. In agricultural areas, farm owners invested in reforestation. In both cases, first-generation planting was often accompanied by slash-and-burn agriculture. This saved work in site preparation and weeding, provided forestry workers with food, and saved planting costs by maintaining a tenancy relationship with the workers. In this slash-and-burn agriculture, foxtail millet, barnyard grass millet, buckwheat, and turnips were produced in the first year. In subsequent years, planting was done and cash crops such as foxtail millet, barnyard grass millet, soybeans, azuki beans, and taro were grown for the next two to five years.

Second, many of the advanced forestry areas developed in the regions surrounding the big cities of Tokyo and Osaka, and each area generally produced a specialty product. Examples include wood from Sanbu for ship construction, or sliding doors and paper screens for Japanese-style rooms; cooperage wood from Yoshino; wood from Tenryu for common lumber; pillar and scaffolding-pole wood from Nishikawa and Owase; decorative pillars from Kitayama; cooperage and utility poles from Chizu; wood from Kito for siding and shutters; and wood from Obi for ship construction.

Finally, silvicultural techniques used in these plantation forests differed from place to place. For example, the planting density ranged widely, from 1,000 seedlings per hectare in Obi to 10,000 seedlings per hectare in Yoshino, and cutting rotations varied from 30 years to more than 100 years, according to local custom and product specialty. Consequently, the forest management practices employed also varied by region. Pruning was introduced in Nishikawa, Owase, Kitayama, and Chizu, while improvement cuts of poorly formed trees in immature stands (Jobatsu) was carried out in Nishikawa, Owase, and Kitayama. Thinning was done in Owase, Yoshino, and Kitayama. Thus, different silvicultural techniques were formulated in different regions, according to the natural local conditions and prevailing market situations.

National forest land that was considered to be expendable was sold in the early 20th century, and the resulting income was used to initiate a special management program that was implemented over the next 20 years. This was the beginning of plantation forest management in the national forests. Initial planting density was 5,000 seedlings per hectare, but gradually decreased to about 4,000 seedlings per hectare. Later, however, reforestation in the national forests decreased dramatically due to funding shortages and instances of unsuccessful plantations. There were also experiments with selection methods based on the concept of sustained yield. These were primarily focused on regeneration by natural seeding, and no new silvicultural techniques were developed.

Reforestation Policy after the Second World War

Postwar reforestation began with the need to restore those forests that had been harvested to provide wood for wartime military use. This restorative reforestation ended in 1956, when the Economic White Paper declared that Japanese war damage had been repaired and that the economy was growing rapidly. Subsequently, in an effort to transform slow-growing natural forests into fast-growing coniferous forests, large-scale afforestation was begun in both national and private forests.

The Forestry Agency established a National Forest Long-Term Production Plan in 1953. Aimed at producing coniferous timber, silviculture was based on the clear-cutting method while the practice of any kind of selection method was limited. Clear-cutting was gradually accepted during the process of salvage logging and the restoration of damaged land after a major typhoon in 1954.

Two factors directed national forest silviculture towards the development of plantation forests. One was the necessity to increase national income in accordance with the introduction of the independent accounting system. The second was the need to meet the increasing demand for wood during a period of high economic growth. In the mid-1950s, short-rotation forestry was also considered. It was aimed at small-log production by: (1) promoting tree breeding through superior tree selection and cloning, (2) introducing foreign species such as *Eucalyptus globulus* and *Pinus taeda,* (3) introducing Japanese larch (*Larix kaempferi*) to remote mountain areas and cool-temperate zones, (4) applying fertilizer to forestland, and (5) introducing terrace planting to promote mechanization.

Agroforestry was also considered, and a forest soil survey was conducted for the "right tree on right site" forestry program. In the late 1950s, short-rotation silvicultural techniques were tried with conifers such as sugi and *Pinus* (20-30 years) and hardwoods such as *Acacia* and *Alnus hirsuta* (10-20

Figure 2.2

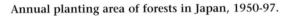

Annual planting area of forests in Japan, 1950-97.

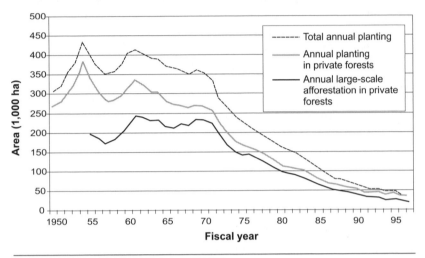

Source: Forestry Agency, *Ringyo tokei yoran*, 1952-99.

years). Based on the experimental results, the Timber Production Increase Plan, introduced in 1961, intended to increase plantation area by 2-5% and harvest volume by 20% above the National Forest Productivity Increase Plan (established in 1957). Although the Timber Production Increase Plan intended to increase the planting density of sugi and hinoki in the national forests from the standard of 3,500 seedlings per hectare to 4,500, it was not fully realized. Actual results for sugi were 3,500 and for hinoki 4,000 seedlings per hectare.

Private forestry economics changed drastically, however. The two biggest changes were the so-called fuel revolution, a shift from fuelwood to petroleum, and the demand for urban housing created by growing population concentrations in big cities. With the fuel revolution, the demand for hardwood fuel declined. The housing construction boom rapidly increased the demand for coniferous construction timber, however, causing prices to skyrocket. Both of these factors encouraged a change in plantation species from hardwoods to conifers. Also, since hardwoods replaced conifers for pulp, it was possible to sell the hardwoods, which were cut before conifers were planted, as pulpwood.

Because it was easy for forest owners to earn the money to invest in planting, this became the driving force behind large-scale afforestation. In 1961, at the peak of this large-scale afforestation, 240,000 ha were planted (Figure 2.2). There was, however, a significant problem looming. Since most of the planting was done over a short period by inexperienced farmers, the

Figure 2.3

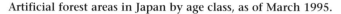

Artificial forest areas in Japan by age class, as of March 1995.

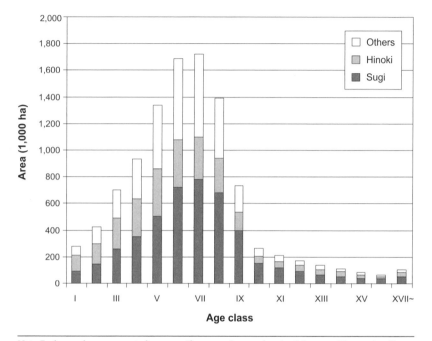

Note: Each age class represents five years (for example, age class I = 1-5 years old, age class II = 6-10 years old, etc.).
Source: Forestry Agency, 1996.

technical silvicultural instructions provided by public-sector forestry pro-motion offices played an important role, resulting in the nationwide adop-tion of the silvicultural techniques prescribed in the manual in connection with expansion of the planting subsidy. A concentration of plantation for-ests of the same stand, structure, and age class resulted (Figure 2.3), later causing a nationwide maintenance problem.

After wood imports were liberalized in 1960, the volume of wood im-ported increased faster than expected. This eased the tight supply situation and caused wood prices to decline. Large-scale afforestation was no longer attractive to forest owners, so there was no reason to further promote it. Some plantations were also unsuccessful because trees were planted on un-suitable sites or the wrong species were selected. At the same time, there was an increasing social expectation that forests should fulfill an environ-mental function, a reaction to some of the consequences of rapid economic growth, such as air and water pollution. Finally, in the early 1970s, the widespread use of clear-cutting for increased timber production was reviewed and the rate of reforestation subsequently decreased (Figure 2.2). In 1973, the Forestry Agency declared a "New Forest Management" policy for the national forests, which emphasized their environmental function and encouraged a shift to small-scale clear-cutting or selective cutting.

Labour-Intensive Silviculture and High-Quality Log Production

In the late 1960s, as wood prices stabilized with the further liberalization of wood imports, the price differences between species and grades of lumber increased. This was because regular-quality domestic lumber had to seriously compete with imports, whereas there was a niche market for high-quality lumber with the decorative attributes demanded by consumers. Structural changes in the demand for wood, however, also influenced lumber price differences.

First, there were price differences between species. High economic growth rates encouraged investment activity and rapid population growth in cities, creating a big demand for uniform-quality construction lumber. Fire prevention regulations, however, limited the use of wood in some parts of buildings, such as outer walls in densely populated areas, thereby decreasing demand for wood products. Responding to this demand shift, large quantities of imported lumber and non-wood products were supplied to the market, since by pursuing economies of scale they could be produced cheaply and consistently. (They were used for sills, studs, panel studs, beams, and small-sized lumber.)

The domestic sawmill industry found it difficult to compete in the market, let alone influence price levels. The situation was especially severe for sugi and *Pinus* lumber, which often competed with imported and non-wood products. Hinoki, for which Japanese consumers are said to have a special preference, was much less affected, however. This was the biggest factor in expanding the price differential between hinoki and other species. The prolonged price stagnation for sugi and *Pinus*, the dominant plantation species, discouraged forest owners from harvesting timber, investing in silvicultural activities, or improving their silvicultural techniques.

Second were the price differences between log and lumber grades. Population concentration in cities led to a great housing demand, causing land prices to skyrocket. Consequently, large numbers of new, suburban houses were built, many of which were apartments or tiny houses, sometimes called "rabbit hutches," whose standardized designs used more or less the same materials. Reflecting a more Westernized lifestyle, most houses had only one or two Japanese-style rooms, typically used as guestrooms. By using decorative lumber for fancy fixtures such as posts, sills, and door lintels in the Japanese-style rooms, housing companies and carpenters introduced a marketing strategy of providing differences and personality to each house.

This strategy met the demands of many consumers who preferred Japanese-style rooms and wanted their houses to have some individuality. The decorative lumber used had to have few or no knots, beautiful colour, and attractive annual rings. These factors led to price differences between grades, and the degree to which lumber satisfied these quality standards affected its price.

Logs that could be processed into decorative lumber were called quality logs, and forest owners became interested in quality-log production. Most quality logs were supplied by advanced forestry areas because it was difficult for the typical postwar forestry areas to produce them. Notable among the advanced forestry areas were Yoshino and Kitayama. Yoshino had established long-rotation forestry, characterized by dense planting and frequent thinning. About 10,000 sugi and hinoki seedlings were planted per hectare, and a variety of high-quality logs, ranging from small to large, were produced during the long rotation. Alternatively, Kitayama was well known for short-rotation forestry of sugi, in which rooted cuttings were used and the trees were pruned, for both knotless log production and plantation density control. Kitayama logs were not sawn, just peeled and polished to make fancy wooden posts. In both forestry areas, systematic silvicultural methods were practised to achieve specific production goals.

Taking Yoshino and Kitayama as their models, forest owners from other areas who were interested in quality-log production recognized the necessity of formulating silvicultural systems suitable for each local area. Many such models were formulated based on stand density control diagrams made by research institutes and administrative forestry promotion sections. Experimental silviculture by individual forest owners or by forest owners' study groups was also incorporated. Most models resulted from minor changes to, or mixtures of, the Yoshino and Kitayama methods; thus, no original models were developed from silvicultural systems adapted to local conditions. In many forestry areas, since most of the plantations were 10-20 years old, when pruning was due, only the pruning operations were actively practised for quality-log production. In those days, since the best-established pruning technique was from Kitayama, it was naturally the Kitayama method that was widely applied. The Kitayama method was developed to produce fancy polished logs, however, not lumber. In some areas, the Kitayama method caused black spots in the heartwood, rendering the logs unsuitable for decorative lumber. When the black colouring appeared on the sawn surface, the market prices of these pruned logs dropped considerably. Although systematic silvicultural practices were required for quality-log production, the only silvicultural activity actually introduced was pruning. In some cases, additional investment in pruning practices even lowered the profitability of forestry.

Increases in Silvicultural Costs and the Stagnation of Silviculture

In the late 1970s, imported wood accounted for more than half of the total wood supply and had a predominant influence on price levels. Because of the shift from a fixed to a floating exchange rate system, appreciation of the yen continued and wood prices began to collapse. Economic growth increased wage rates, further reducing forestry profitability. In the 1980s, the market share of non-wood and wood houses built using two-by-four or pre-fabricated methods increased. This resulted in a decrease in traditional post-and-beam housing, which uses more domestic wood than the other construction methods. Even the traditional houses had fewer Japanese-style rooms. Since the late 1980s, large housing companies have been using considerable amounts of glue-laminated wood for fancy fixtures. All of these construction changes have depressed the demand for high-quality logs.

By this time, most plantation stands had grown to the point of needing improvement cuts and thinning. In Kyushu and some areas of Shikoku, harvests were planned before other areas, and, using lightweight skyline and mobile carriages, a small-scale hauling system was constructed to transport the thinned logs. Thinning subsidies were also increased.

For several reasons, however, thinning was not carried out as actively as expected. First, thinning profitability was too low to create direct income. Second, because of the low returns from forestry, many forestry-oriented households developed income sources other than forestry or agriculture, leaving less family labour available for forestry work than previously. Third, the demand for quality logs, which had previously been an important incentive for forest owners to invest in silviculture, declined, especially recently. Because many forestry households lost interest in silvicultural investment, the delay in thinning plantation forests became a big problem.

To counter this decline in both forest owners' investment and harvest volume, the government introduced two policies: one encouraged organization of local forestry and the other reviewed silvicultural systems. The first policy attempted to improve market conditions for domestic wood by reorganizing the processing and distribution sectors. In hopes of raising wood supply capability, the policy also promoted cooperation in silviculture and log production through activity led by forestry cooperatives. This concept was first published in the Forestry White Papers of 1976 and 1979. This policy's purpose was only the commercialization of wood from plantation forests, not forest productivity expansion or the development of a new silvicultural system.

The second policy, reviewing silvicultural systems, began in 1987 and was published in that year's Forestry White Paper. The main idea was "low-cost oriented forestry techniques," achieved by introducing high-tech machinery and, when possible, utilizing the forest's natural forces. Located in the temperate monsoon climate zone, Japan has a climate that is mild with

Table 2.1

Labour input for different levels of silviculture in sugi forests in Japan in 1981 and 1991.

Activity	Labour input (person-days)	
	1981	1991
Site preparation	20	14
Planting	17	14
Supplementary planting	4	2
Weeding	80	75
Raising fallen trees	13	12
Cleaning	15	10
Pruning	23	24
Other work	15	6
Total	**187**	**157**

Source: Ministry of Agriculture, Forestry, and Fisheries, Ikurinhe Chosa Hokoku, 1983, 1993.

abundant rain, and therefore suitable for vegetation growth. Hence, in the silviculture of plantation forests, good care of forestland by site preparation, planting, weeding, and cleaning is usually indispensable. This, however, pushes Japanese silvicultural costs to higher levels than in other countries. By introducing a technique utilizing the forest's natural forces, the new policy tried to reduce the labour required for site preparation, planting, weeding, and cleaning. A target was set to halve the labour necessary for one clear-cutting rotation from 200 worker-days (w-d) per hectare (ha) to 100 w-d/ha, thereby also halving the per-hectare silvicultural costs from ¥2 million to ¥1 million (from US $16,700 to US $8,300). Various techniques, including pot nurseries, automatic planting machines, tractors, herbicides, and fertilizers, were encouraged.

Table 2.1 compares the labour input for different levels of silviculture in 1981 and 1991. The data are taken from the national average for sugi forests, and are based on silvicultural work actually performed in those years by 400 forestry households that owned more than 20 ha of forest. While undeniably heavily influenced by the individual households' situations and by climatic conditions, the data reflect the general tendency of decreasing labour input into all practices, which during the decade from 1981 to 1991 led to a decrease of 30 w-d of total labour input. This decrease did not necessarily result from labour-saving techniques, but rather reflects the fact that forest owners have consistently reduced labour input to the minimum level necessary. Low prices for wood have made the profitability of forestry too low for the cost of labour investment in silvicultural practices to be attractive. In practical terms, labour-saving silviculture has almost precipitated the abandonment of silviculture in Japan today.

The Pursuit of Alternative Silviculture

To try to reverse the decline in Japanese silviculture, three approaches have been proposed: (1) long-rotation forestry, (2) re-evaluation of uneven-aged forests and natural forest management, and (3) agroforestry and urban forestry.

Long-Rotation Forestry

Long-rotation forestry extends the standard rotation period of 50-60 years to longer periods. Behind this idea is the political intent to spread out cutting times, thereby correcting the age-class distribution imbalance in Japanese forests. As already noted, most reforestation occurred in the 1950s and early 1960s. Age classes, therefore, are unevenly balanced, which is not good for future timber production (Figure 2.3). Regardless of the political intent, however, most forests will remain uncut and will progress from middle to old age because forest owners have lost interest in silviculture. Such promotion of long-rotation forestry equates to "approved" abandonment of silviculture. If forests are simply left to age without being tended, they will not produce good-quality timber, nor will they perform the environmental function expected of well-maintained, mature forests.

Re-evaluation of Uneven-Aged Forests and Natural Forest Management

Mixed-age forests and effective selection methods were considered a part of a new silvicultural movement to help maintain forest health and reduce silvicultural costs. This was not a completely new idea, having already been tried in several places. Examples in private forests are Imazu (Gifu Prefecture), Kuma (Ehime Prefecture), Noto (Ishikawa Prefecture), and Tane (Shiga Prefecture). In most of these areas, however, either selection methods have been abandoned or mixed-aged forests are gradually becoming even-aged, as mature trees are cut without subsequent reforestation. Only in Kuma forest have some individual forest owners persisted, with national university and research institute support. The overall lack of success is attributed to several factors: (1) no existing economic base to support the uneven-aged system, (2) no existing silvicultural manual, and (3) few workers sufficiently trained in selection. Thus, despite policy encouragement, the forest area managed under the selection method has barely increased.

The government has also promoted the silvicultural management of natural forests, which are composed mainly of hardwoods. The purposes were to reduce silvicultural costs, improve the forest environment, and extend hardwood resources, which are projected to be in short supply worldwide. Except for naturally regenerated hardwood forests for firewood or mushroom production, silvicultural management of privately owned natural forests has only recently been introduced in limited areas. It is therefore more difficult to develop and propagate suitable silvicultural management techniques for natural forests. Research organizations, such as those involved in the national forests in Hokkaido and Kyushu, university forests, and forestry experiment stations have collected data on beech and oak forest silviculture and attempted to develop silvicultural techniques from these data. The problem often identified is that each research project is performed at an individual level and rarely continued at an organizational level. Since the technical basis for formulating new silviculture will not be developed anywhere other than within these research organizations, there is a pressing need to create a research environment that both enables continuous research at an organizational level and facilitates cooperation between the organizations involved.

Agroforestry and Urban Forestry

Recently some forestry researchers have advocated agroforestry and urban forestry. These methods conserve the environment and possibly use the forest's natural forces in various ways. The researchers adopted this approach after learning about the development of silviculture abroad and by reviewing Japanese forestry history. In Japan, only the concepts of agroforestry and urban forestry have been introduced so far. They can, however, provide city dwellers opportunities for relating to forests and participating in forestry work as recreation, thereby reducing silvicultural costs and raising income from different sources, even during early-rotation stages. The introduction of these methods has recently been considered in some university forests, where the results of trials may lead to prospects for new silviculture. Chances are high that these silvicultural methods will be practised in the field and provide positive economic benefits.

Who Should Be Responsible for Forest Management?

Although Japanese plantation forests have been grown as a resource, they will produce suitable wood in the future only if proper thinning is performed. It is important to accept this fact as the starting point for considering future silviculture. Several silvicultural methods to make forests healthier and more abundant have been proposed recently. The problem is that forest owners lack any strong economic incentive for pursuing this new silviculture.

Today, there is worldwide discussion about sustainable forest management practices from an environmental viewpoint. Implementing these practices might lead to higher wood prices, the expectation of which becomes an incentive for forest owners in other countries to adopt silviculture. This will not be the case in Japan, however. Since most forest owners are no longer actively involved in forestry, the economic basis for forest management has already been lost. This situation resulted from (1) the decline or abandonment of agriculture by households that previously also managed plantation forests, and (2) much lower income for forestry workers compared with most other occupations. In general, forestry is no longer seen as an attractive investment asset, and many owners attribute no asset value to the growing forests in which they have invested. Today's forest owners, who are supposed to perform and invest in silviculture, have detached themselves from forestry.

Thus, while society has become more interested in silviculture, forest owners have lost interest. Strong will and immediate decisive action are needed to change the situation facing silviculture, from the viewpoint of both a shared environment and forest resources worldwide. Since it is difficult for the private sector to lead this action, the public sector (e.g., local municipalities) must take the initiative. These public organizations have sufficient experience, information, and familiarity with the local environment to direct silvicultural development by reorganizing the whole process, from wood production to distribution to processing. They are likely to propose silviculture suitable for their local forests because they have direct incentives to invest in the very forests that comprise their own environment. Organizations in the local public sector also have the responsibility for fair distribution of goods and services resulting from silvicultural practices. As for the forestry workforce, it could be organized either by forestry cooperatives or by the semi-public sectors to which a greater variety of subsidies are available. Whichever path is followed, the core of the workforce should be young people employed under modern hiring conditions.

In promoting an initiative by local municipalities, the national government hammered out the Municipal Forest Plan in the 1990s. Under this plan, municipalities are responsible for developing a forest improvement plan for private forests in their region. These plans include thinning and tending, the promotion of cooperative silvicultural practices, and the recruiting and training of forest workers. The plans can also specify which forests need thinning and tending urgently, and how and when these should be done. If a forest is left untended despite being recognized as being in need of thinning and tending according to the plan, the municipal authority can advise the forest owner to undertake the thinning and tending required. Further administrative guidance is offered if this advice is not followed; if all else fails, a prefectural governor could even rule that the forest owners must make a profit-sharing contract with the municipality. This approach reflects the political intent to establish a system of organized, collective silviculture, instead of leaving it to individual forest owners' decisions and actions. Today's problems of managing plantation forests are the result of large-scale postwar afforestation policies. Policy now focuses on collective silviculture, beyond the interests of individual forest owners. There has, however, been insufficient discussion of the ways and the extent of the transfer of silvicultural practices from the private to the public sector, and further discussions are anticipated. In practice, no municipal governor has ever advised a forest owner to practice thinning and tending, and the plan is backed with insufficient funds. These limitations show that the idea of transferring silvicultural practices from the private to the public sector is not yet widely recognized or accepted by society. There is an urgent need to engage in discussion, reach agreement, and take action wherever possible.

References

Handa, Ryoichi, and Morita, Manabu, eds. (1979). *Nihon Ringyo no Shinro wo Saguru (I)*. 315 pp. Nihon Ringyo Kyokai, Tokyo (in Japanese).

Kurasawa, Hiroshi, ed. (1965). *Ringyo Kihonho no Rikai*. 378 pp. Nihon Ringyo Chosakai, Tokyo (in Japanese).

Murao, Koichi (1969). *Ikurin no Seisan Kozo*. 257 pp. Rinya Kousaikai, Tokyo (in Japanese).

Nihon Ringyo Gijutsu Kyokai (1972). *Ringyo Gijutsushi Volume 1*. 727 pp. Nihon Ringyo Gijutsu Kyokai, Tokyo (in Japanese).

– (1976). *Ringyo Gijutsushi Volume 2*. 667 pp. Nihon Ringyo Gijutsu Kyokai, Tokyo (in Japanese).

Rinsei Sogo Kyogikai (1980). *Nihon no Zourin Hyakunenshi*. 425 pp. Nihon Ringyo Chosakai, Tokyo (in Japanese).

Shinrin Keikaku Kenkyukai (1987). *Aratana Shinrin Ringyo no Choki Vision*. 415 pp. Chikyusha, Tokyo (in Japanese).

3
Private Forestry
Ken-ichi Akao

This chapter deals with private forestry in Japan. Most statistics define a private forest as a privately operated forest, and the statistical figures in this chapter use this definition. In practice, most private forests are managed by the owners, therefore no strict distinction will be made between ownership and management unless stated.

This chapter is structured as follows: The first section gives an overview of private forestry in Japan. The second section outlines the present state of forest resources as well as changes since the Second World War. Changes are characterized by an increase in growing stock and promoting the conversion from natural to planted forests. The third section focuses on reforestation, while the fourth investigates the increase in growing stock in light of harvesting activity, which had declined and remained stagnant since the 1960s. The fifth section reviews the related forest policies of the central government. The sixth section describes the response of forest managers to the severe market conditions of the last two decades. The last section presents the perspective of private forest management in Japan.

Overview of Private Forestry in Japan
As shown in Table 3.1, participants in private forestry are classified into six categories: forestry households, companies, shrines and temples, joint holdings, various groups and cooperatives, and habitual joint holdings. Except for shrines and temples, these groups manage forests to produce income. In general, these forests are not very large: the average forest area per establishment was 3.7 ha in 1990. Groups operating forests of less than 100 ha control 67% of the total area. Hence forests are not necessarily owned by a few large concerns. One characteristic of private forestry in Japan is this dominance of small management with respect to both number and area. This is particularly evident for forestry households, which have a 90% share in number and 63% share in area. This relates closely to the history of private forest ownership.

Private ownership of forestland was officially permitted after the Meiji restoration of 1868, which marked the end of feudalism. Previously, forestland (excluding land owned by central and local governments) was legally owned collectively for the use of local communities. After the Meiji restoration, most of these "commons" became privately owned land, although some still exist and are operated by permanent joint holdings. In many cases, common forests were divided equally among members of the community, and individual parcels of forestland were small. This is the origin of forests operated by forestry households. A large part of forests managed by other groups also originated in the former commons. Although there are a few forestry households with large areas of nearly 1,000 ha, without exception they have all bought numerous small and geographically scattered forest holdings over a long period.

A few companies acquired forest areas in other ways. Some papermaking and metal-refining companies acquired much of their forestland from the central government in the early 20th century. At that time, the government disposed of a part of the national forests in order to raise funds for public works and to promote modern industries such as papermaking and metal refining. This is the origin of large forests owned and managed by companies. These cases are rare, however, and in terms of average figures, forest areas operated by companies are not necessarily large. In particular, forest area per company decreased markedly in the period between 1960 and 1990. This is why firms not directly related to forestry, such as real estate agents and service industries, tended to actively purchase forestland. In many cases, their aim was not forest management but ownership of the forest as an asset in their portfolio and/or utilization of the land for some other use, such as housing developments or golf courses.

Table 3.1 shows that in the three decades between 1960 and 1990, the average private forest area hardly increased. While this tendency might have been evident in earlier periods, the relevant data are not available. There are, however, three reasons why the scale of management has not increased.

First, it is not a Japanese trait to willingly convert the ownership of real estate, including forestland, into other assets such as monetary instruments. When the monetary and insurance markets were immature, particularly before the Second World War, forestland was one of the few investments available. In addition, Japan had experienced several periods of inflation accompanying economic growth, which contributed to a strong preference for real estate as an investment strategy.

The second reason concerns an aspect of silvicultural technology. Standard silviculture in Japan does not require a strict time schedule for forest operation. For instance, it is advisable to weed (essential for most silviculture in Japan) before summer. If the work is delayed, however, or, except for the first four or five years after planting, even if the work is not performed

Table 3.1

Forest establishments and forest area in Japan in 1960 and 1990.

Year	Establishment category[1]	Number of establishments			Size of establishments[2] (1,000 ha)			Average establishment size[3]
		Total	≥100 ha	% ≥100 ha	Total	≥100 ha	% ≥100 ha	
1960	Forestry households	2,705,269	2,927	0.1	6,350	738	11.6	2.3
	Companies	3,303	330	10.0	715	658	92.1	216.3
	Shrines and temples	20,212	50	0.2	73	25	34.6	3.6
	Joint holdings	148,515	558	0.4	544	149	27.3	3.7
	Various groups and cooperatives	2,918	140	4.8	73	51	69.4	25.0
	Habitual joint holdings[4]	109,909	2,793	2.5	1,567	1,015	64.8	14.3
	Total	2,990,126	6,798	0.2	9,321	2,635	28.3	3.1
1990	Forestry households	2,508,605	3,753	0.1	6,752	1,052	15.6	2.7
	Companies	43,937	1,372	3.1	1,520	1,345	88.5	34.6
	Shrines and temples	33,628	155	0.5	153	64	41.8	4.5
	Joint holdings	202,786	712	0.4	710	199	28.0	3.5
	Various groups and cooperatives	11,327	754	6.7	405	278	68.7	35.8
	Habitual joint holdings[4]	59,209	2,265	3.8	1,142	684	59.9	19.3
	Total	2,859,492	9,011	0.3	10,683	3,623	33.9	3.7

1 All figures are for establishments operating forests greater than or equal to 0.1 ha.
2 The compilation of area is based on data from self-enumeration by each establishment. They often state the *registered* forest area found in the register book, not the *actual* area. In many cases, there is significant difference between the registered and the actual area. It is not unusual that the actual area is three times larger than the registered area. More accurate figures for forest area appear in Table 3.2, but it is not possible to yield figures consistent with Table 3.2 and classified by each establishment categorized here.
3 Hectares per establishment.
4 "Habitual joint holdings" denotes that the land is owned as local common property, a relic of ancient ownership.
Source: Ministry of Agriculture and Forestry, *The Forestry Enterprise Survey for the World Census, 1960 and 1990*.

over a few years, harvesting revenue is not drastically reduced. This flexibility enables forest owners to have another job and manage their forests part-time. Forest owners therefore did not have a strong incentive to entrust their forests to professional managers. The potential demand for professional management services appears to have grown, however, since the number of owners living far away from their forests has increased and these owners do not have related forest work experience.

The third reason arises from the attitude of forest managers. The advantage of scale in forestry can be realized when the manager operates a sizable forest unit. More precisely, when managers have large forest units, construct an effective forest road network, and obtain high-performance forestry machinery, they can achieve advantages of scale. However, it is difficult to obtain large forest units because of the two reasons outlined above. Furthermore, most forests in Japan are located on steep land in mountainous regions, and it is extremely costly – an average of several hundred dollars per metre – to build forest roads. Enlarging existing forests was therefore neither easy nor economically attractive to forest managers.

Given the difficulty of enlarging existing forests, forest managers have concentrated on the labour-intensive technologies available for forests. On the other hand, they were less interested in labour-saving (capital-intensive) technologies (e.g., road construction). In 1996 the total length of forest roads, excluding national forests, was 86,000 km, and the average length was 5.0 m/ha. As a result, forestry labour productivity remained quite low. Minamikata (1995) estimated the average labour productivity of log production for Japan, Sweden, and Canada in the early 1980s. The productivity was 2.0 m^3 per worker-day in Japan, compared with 7.1 in Sweden and 13.2 in Canada.

Generally speaking, in a developing economy, wage rates or opportunity costs for labour continue to increase, and in fact the wage rates of forest workers in Japan have been increasing (Figure 3.1). This causes industries that do not adopt labour-saving technologies to decline. It is this problem that has been facing private forestry in Japan for the past two or three decades.

Forest Resources
Table 3.2 shows the present state and changes in the forest resource in Japan. According to the 1995 assessment by the Forestry Agency, the private forest area was 14.6 million ha, constituting 57.9% of the total forest area. The high percentage of private forest is one of the salient characteristics of Japanese forestry. It is also evident that this area scarcely changed between 1965 and 1995. Although part of the private forests has been converted into agricultural land, golf courses, residential land, and other developments, some agricultural land has been converted back into forest because of the

Figure 3.1

Wage rates of forest workers in Japan, 1956-96.

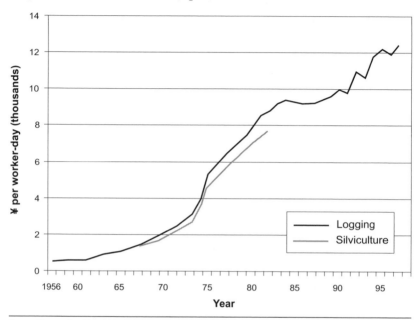

Source: Somu-cho Tokei-Kyoku, *Japan Statistical Year Book* (annual issues); compilation by Ministry of Agriculture and Forestry.

decline in agriculture, and the changes have tended to cancel each other out. Private forest area increased between 1946 and 1965, but the figures for 1946 in Table 3.2 are not comparable with data after 1965 because reliable forest assessment, including classification by ownership, began only in the 1960s.

In Japan, artificial forests are defined as planted forests where the planted trees survive at a ratio of 50% or higher. On the other hand, natural forest is defined as forest other than artificial forest, including virgin forest and secondary forest. Table 3.2 shows that the share of artificial forest in private forests has been higher than in the total forest area since 1946. This suggests that private sector planting has been more active than public sector planting in the prewar and postwar periods. From Table 3.2, it can be calculated that the average annual planting rate in private forests was 35,000 ha per year for 1868-1946, 132,000 ha for 1947-65, 77,000 ha for 1966-80, 27,000 ha for 1981-90, and 14,000 ha for 1990-95. It is assumed that the forest area in 1868 was zero, which overstates the annual rates for 1868-1946. While reforestation from natural to artificial forests proceeded rapidly during the first two decades of the postwar period, planting activity subsequently declined, and recently there has been less reforestation activity than in the prewar period.

Table 3.2

Area and growing stock of private forests in Japan, 1946-95.

Forest	Year	Total Area[1]	Total Growing stock[2]	Artificial forests Area[1]	Artificial forests Growing stock/ha[2]	Natural forests Area[1]	Natural forests Growing stock/ha[2]	Bamboo area[1]	Other area[3]	% Artificial forest
Private forest	1946[4]	9,793	(na)[5]	2,746	(na)	5,372	(na)	119	1,556	28.0
	1965	14,193	55	5,245	67	8,336	51	180	432	37.0
	1980	14,733	98	6,399	124	7,749	85	140	445	43.4
	1990	14,651	134	6,673	180	7,393	104	147	439	45.5
	1995	14,572	152	6,743	208	7,211	112	147	471	46.3
Japan[6]	1946[4]	48.0	(na)	62.1	(na)	42.7	(na)	94.9	47.7	21.7
	1965	56.5	75	68.4	71	52.8	85	98.4	29.4	30.5
	1980	58.3	98	64.7	107	55.4	102	97.2	35.7	39.1
	1990	58.1	124	64.6	155	54.7	114	97.4	36.3	41.0
	1995	57.9	139	64.8	182	53.9	119	96.7	38.8	41.4

1 In thousands of hectares.
2 In millions of cubic metres per hectare.
3 "Other area" includes wetland, scrub, shrub, and bushland.
4 The figures for 1946 are not comparable with other figures because of changes in the assessment system.
5 (na) = "not available."
6 For area, the value represents percentage share of private forests.

Sources: Somu-cho Tokei-kyoku, *Nihon Toukei Nenkan* (Statistics Bureau, Management and Coordination Agency, Government of Japan, *Japan Statistical Yearbook*), annual issues (compilation by Ministry of Agriculture and Forestry for 1946; by Forestry Agency for 1965-95).

Regarding growing stock, private forests have become more abundant in terms of volume per hectare than public forests since 1980. This indicates that, for some time, more private forests have begun to be harvested over longer rotations than public forests, at least since 1980. Growing stock is increasing in both artificial and natural forests, and the tendency is marked in artificial forests. The average annual incremental rate of growing stock for artificial forests was 3.8 m³/ha for 1965-80, 5.6 m³/ha for 1980-90, and 5.6 m³/ha for 1990-95.

In summary, forest managers in Japan aggressively converted natural forest to artificial forest during the first two decades of the postwar period, and thereafter have maintained forest resources so that growing stock has increased rapidly over the last three decades. The next two sections consider this trend in detail.

Reforestation

According to annual data published by the Forestry Agency, the area reforested from natural to artificial forest constituted more than 150,000 ha annually for private forestry between 1960 and 1970. This means that more than 1% of private forests were cut and planted every year during the decade. Figure 3.2 shows, however, that the annual reforestation area has been

Figure 3.2

Reforestation from natural to artificial forest in Japan, 1955-97.

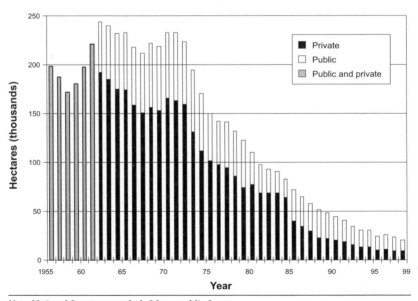

Note: National forests are excluded from public forests.
Source: Forestry Agency, *Ringyo hakusho* (annual issues).

decreasing since at least 1961 (data prior to 1961 are unavailable). In 1997 the area was about 1/20 of the area in 1961 (9,023 ha in 1997 compared with 191,952 ha in 1961), a drastic decline.

Since 1955 the Forestry Agency has monitored the annual numbers of seedlings by species planted on forestland excluding national forest. According to the data, the major species have been sugi (*Cryptomeria japonica*) and hinoki (*Chamaecyparis obtusa*), representing 63% in 1955 and 73% in 1993 and maintaining a rate of over 60% during this period. Hinoki tended to increase (from 20% in 1955 to 46% in 1993), whereas sugi tended to decrease (from 43% in 1955 to 27% in 1993).

Both are coniferous species whose timber is sawn into housing materials. On the other hand, wood produced from the natural forest was used mainly as fuel and charcoal before 1960; since 1960 most has been utilized as pulpwood. This change in usage accompanied the so-called fuel revolution in Japan.

In order to consider changes in reforestation activities in private forestry during the postwar period, the movement of stumpage prices needs to be examined. Three trends can be observed in Figure 3.3:

1   In all periods there was an obvious gap in price between sugi and hinoki artificial forest stands and natural forest stands. This is the basic reason

Figure 3.3

Stumpage prices for sugi, hinoki, and fuelwoods, 1950-98.

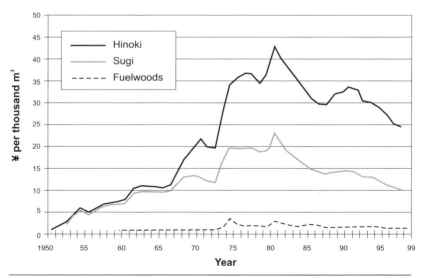

Source: Nihon Fudosan Kenkyu-sho, *Sanrin-soji Oyobi Yamamoto Ryuboku Kakaku Shirabe* (annual issues).

why reforestation from natural to artificial forest proceeded rapidly in the period shortly after the Second World War.

2   The price difference between sugi and hinoki has continued to widen since 1965 as a result of imports of foreign timber that began in the same period. In particular, the price difference resulted from timber imports from North America, which were cheap and closely resembled sugi. In response, Japanese forest managers came to prefer hinoki to sugi for planting.

3   Finally, stumpage prices for both sugi and hinoki began to fall after 1980, whereas previously they had risen constantly. It might be suggested that this decline led to a decline in reforestation activity. It is evident, however, that there was a time lag between the two. Reforestation activity began declining about 1960, and stumpage prices in 1980.

In order to explain the decline in reforestation, other market data are required besides stumpage prices. Theoretical models of forest management such as the Faustmann formula suggest a relative stumpage price, which is stumpage price divided by wage rate. As shown in Figure 3.4, the relative stumpage price has continued to fall more or less consistently since 1960.

Figure 3.4

Relative stumpage prices (stumpage price divided by wage rate) in Japan, 1956-96.

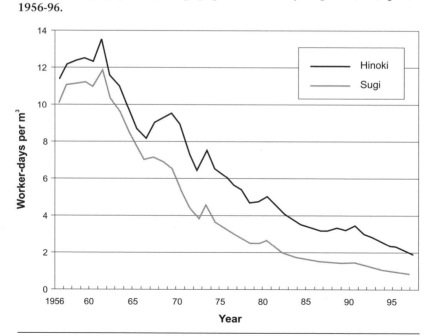

Note: Prices calculated from data in Figures 3.1 and 3.3.

This fact is consistent with reforestation activity observed in Figure 3.2. The decline of relative stumpage prices implies that the cost of labour was increasing faster than stumpage prices and therefore the profitability of forestry declined.

It can be asserted that relative stumpage prices accurately explain the movement in reforestation activity in private forestry in Japan. This indicates that Japanese forest managers tend to be responsive to market conditions. In other words, they are economically rational, contrary to the popular view that forestry households respond very slowly to changing economic circumstances, a position often asserted in the literature on forest economics in Japan. (See the review by Sato [1999] for some typical views on the forestry household held by forest economists in Japan.)

Harvesting

During the postwar period, more than half of domestic log production has been from private forests. In particular, before 1960, when the import of timber was strictly limited because of the shortage of foreign reserves, the share exceeded 80%. Private forestry played an important role in satisfying Japan's timber demand before 1969, the first time that self-sufficiency in timber fell to less than 50% (currently it is less than 20%).

On the other hand, as shown in Figure 3.5, the volume of timber harvested clearly declined from 1957 to the mid-1970s, and thereafter remained at a level of 20 million m^3, about half the production level of the 1950s. It is thought that the harvest decline occurred in both artificial and natural forests. Although detailed data to verify this are not available, this supposition is supported indirectly by the fact that the growing stock continued increasing in both forest types. For sugi and hinoki, the cutting ages were generally thought to be about 40 and 50 years, respectively. If this was still true, log production from artificial forests might have increased in recent years, since a great number of trees were planted early in the decade after the Second World War and would have already reached cutting age. A few studies have reported that log production has certainly increased in recent years in some regions, but this trend has not yet been confirmed by aggregate data across all regions of Japan.

The decline and stagnation in timber harvest volume may well be explained by the idea that Japanese forest managers behave rationally from an economic perspective. Comparative statistics (sensitivity analysis) of deterministic single-stand models indicates that as relative stumpage price declines, optimal harvesting ages become higher. The claim holds true when the manager adopts any of the Faustmann rule, the net maximum sustained yield rule, or the maximum internal rate of return rule (Johansson and Löfgren 1985, chapter 4). The result also holds true in the associated stochastic model for the Faustmann rule (Akao 1996). As shown in Figure

Figure 3.5

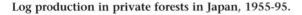

Log production in private forests in Japan, 1955-95.

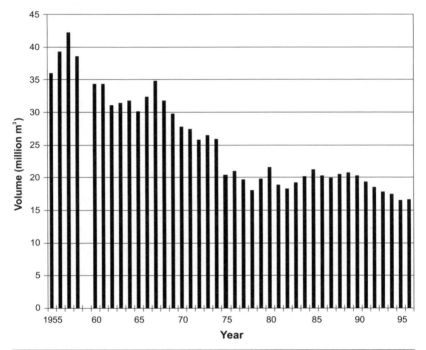

Note: Data for 1959 are unavailable.
Source: Ministry of Agriculture and Forestry, *Mokuzai jukyu hokokusho* (annual issues).

3.4, the relative stumpage price has consistently declined since 1960. Hence it seems natural for Japanese forest managers to postpone harvesting in response to market conditions. The optimal harvest age calculated in recent empirical research is about 80 years for 70% of the surveyed 248 forest units, 45 for 30% of the forest units of sugi, and concentrated around 54 years for the surveyed 396 forest units of hinoki (Akao 1993). Thus, given current wage levels, we would not expect to see increased timber harvest volumes, at least during this decade.

Forest Policies in Relation to Private Management

So far we have discussed a number of significant topics in private forestry in the postwar period and related them to changes in market conditions. In this section, various central government policies related to private forestry are examined.

Since 1868, when the Meiji restoration occurred and Japan established a decentralized market economy, the national goal for forests and forestry has been to build up the forest resource rapidly and satisfy the demand for forest products by utilizing domestic resources. In order to accomplish this, the central government has treated private forestry sympathetically through several policy instruments.

In the early postwar period, the government subsidized half the planting cost in order to rapidly reforest areas destroyed in the war. Afforestation on bare land was completed within a decade, and then reforestation from natural to artificial forest came to be regarded as important. After 1957, 48% of the planting cost for such reforestation was provided as a subsidy by the central government, compared with 24% for the replanting of artificial forests. The amount of subsidy has changed several times, but the relative difference has been maintained up to the present.

Favourable loans are also available for forest management. The Agriculture, Forestry and Fisheries Finance Corporation, an affiliated group of the central government established in 1953, provides funds for forestry at fixed interest rates that are lower than prevailing rates from private banks. Repayment terms are over decades and most funds have been provided for planting.

With regard to related tax schemes, households receive income tax relief, whereas corporations do not. For forestry households, only one-fifth of revenue from harvesting is subject to income tax, with the total amount of income tax being derived by multiplying the evaluation by 5. Since income tax is progressive in Japan, this "divide by 5 and multiply by 5" procedure lightens the tax burden for forestry households. There is also favourable valuation of forest assets inherited by forestry households, although it is often observed that the inheritance tax requires too much felling to sustain forest management by forestry households, particularly those that inherited large forests of more than 100 ha.

Obviously this favourable treatment promoted reforestation activity in private forests. On the other hand, the effect on harvesting in artificial forests is not clear. From a theoretical viewpoint, subsidies for planting and favourable income tax treatment promote harvesting by shortening the optimal harvesting age, while providing funds at low interest rates delays the optimal age. Thus the total effect on harvesting cannot be clarified without further information. It does not appear that policy-makers in the past intended to control harvesting activity using the above policy instruments, as their central concern was to promote reforestation from natural to artificial forests and to protect forestry in Japan.

The Forest Plan System begun in 1968 controls private forestry in other ways. It allows local or central governments to approve a forest plan made by a forest manager or managers. The plan is called the Forest Management Plan. When the plan is authorized, favourable consideration is provided to forest operations undertaken according to the plan. For instance, the standard subsidy for planting cost in reforestation is 32%, compared with 48% when the reforestation is conducted following the authorized forest plan.

The Forest Management Plan is a comprehensive plan including planting, silviculture, and harvesting. By authorization and favourable treatment, the government attempts to promote forest management from a long-term rational perspective.

Private Forestry in the Forest Depression

As seen from the simple analysis in the previous sections, since the 1960s market conditions relating to private forestry have become severe. Since 1980, stumpage prices for sugi and hinoki have fallen even in nominal terms. In this section the response and efforts of forest managers under these harsh circumstances are examined.

In any industry, if market conditions deteriorate some establishments leave the industry. Forestry is no exception. In the case of small forests, forestry households decrease their labour input and increase their reliance on other sources of income. Originally, most managed their forests part-time and maintained other jobs. It was therefore easy to reallocate labour input. According to data published by the Forestry Agency, forestry household labour input to forestry has been decreasing constantly since 1971.

As for companies, it is also not difficult to leave forestry and enter other industries because companies have comparatively greater manpower resources and can put together an appropriate management team to enter a new industry. In fact, in the case of many pulp firms and homebuilders owning large forests, most forest products have been acquired from imports since about 1970, and now the dependence on domestic industrial forests is quite insignificant.

Full-time forestry households could not leave forestry and had to confront harsh market conditions. For them finding other sources of income meant changing occupations. It is more difficult for them than for part-time forestry households to adjust their labour input. Also, the management scale of full-time forestry households is not as large as for companies, so it is difficult to leave forestry and enter other industries. The following response of forest managers was typically observed for such full-time forestry households.

When market conditions deteriorate or, more precisely, when relative stumpage prices fall, an appropriate policy would be to adopt labour-saving technology to improve labour productivity. In forestry, this is done by the construction of forest roads and utilization of high-performance machinery. This is costly and difficult to carry out, however, because of factors referred to in the first section of this chapter. Indeed, few forest managers can adopt such a policy. Although many managers recognize that labour-saving technology is very important and have tried to reduce their labour requirements in several ways, what could actually be achieved was quite limited. For example, labour input could be reduced by limiting the frequency of weeding in the first 10 years after planting seedlings from 10 times a year to 8. However, the major response of forest managers was to adopt more labour-intensive technology in the following ways.

Forest managers first noticed that the massive supply of imported timber was changing market conditions. Consequently, their efforts turned towards differentiating their products from imported timber (i.e., product differentiation). They aimed to produce the sawnwood used in traditional Japanese rooms for decoration. In these rooms, lumber is used for its aesthetic visual appeal and the Japanese consumer generally prefers domestic species. Forest managers also aimed to produce forest products that could be produced quickly. Since optimal cutting age became higher as relative stumpage price decreased, thus delaying the revenue from forests, such products became important in order to supplement income, particularly for full-time forestry households living on the income generated from their forests. One such product is timber used for posts, which has a comparatively low cutting age (30-50 years). As a result, the major product forest managers tried to produce was timber made into decorative posts for use in traditional Japanese rooms.

Coping with the forest depression in this way led to more labour-intensive forestry, however. Since the Japanese traditionally prefer wooden goods with fine grain and in good condition, it is necessary to prune branches frequently in order to produce timber for decorative use. Pruning requires intensive, trained labour since inappropriate pruning injures the tree and bruising greatly reduces commodity value. Furthermore, because the aim was to shorten the production period, pruning had to be performed at the earlier stages of silviculture and more frequently than in standard silviculture.

In a situation where the potential to adopt labour-saving technology was limited, it might well have been natural to adopt a strategy of product differentiation and to input more labour into the forest than before. On the other hand, wage rates kept increasing, and after 1980 nominal stumpage prices fell. Conditions were becoming increasingly harsh for labour-intensive forestry. In fact, while pruning rapidly became popular during the 1970s, forest managers seemed less enthusiastic about pruning after 1980. In the 1980s, some managers zoned their forests into two areas, a labour-intensive forestry area and a labour-saving and time-intensive forestry area. In the latter zone, the anticipated harvesting age is quite advanced, occasionally over a hundred years. While adopting labour-saving forestry with long rotation periods might be an economically efficient policy, it creates the more serious problem of how to supplement the inadequate income resulting from the delay in revenue derived from harvesting.

Recently forest managers have been engaging in various experiments. Some managers are trying to develop new forest products. Plants that are used in religious rites and that are made into medicinal materials are grown and harvested. In addition to individual management, collective experimentation is becoming popular through, for example, a forest owners' association. In some cases, managers in villages or village groups cooperate to create forest-related industries such as sawmills, landscape architecture, and housing. In other cases, the activity is connected to comprehensive economic development in the region. It is not yet known whether these innovative projects will be successful in the long term.

A Perspective

It is thought that a forest of more than 100 ha is required for a forestry household to live solely from forestry. As seen in the first section of this chapter, however, few households operate forests of this size. As an overall tendency, therefore, private forestry will continue to decline. In fact, aggregate data concerning reforestation, log production, and labour input to forestry show that these activities are in decline. Unless there is a technological revolution in forestry, or unless market conditions change drastically, this trend will not readily change.

The severity of recent market conditions can also be seen from the internal rates of return for sugi, as calculated by the Forestry Agency. As shown in Table 3.3, the rates recently fell below 1%, although this figure does not take into account the possibility that internal rates of return may be increased by selecting appropriate harvesting ages. Statistical results from analysis, including the choice of cutting ages, are given below.

Table 3.3

Internal rates of return (IRR) for sugi plantations, 1965-93.

Year	IRR[1]
1965	6.3
1970	5.6
1975	4.1
1980	3.4
1985	2.1
1986	1.9
1987	1.7
1988	1.7
1989	1.7
1990	1.7
1991	1.6
1992	1.3
1993	0.9

1 Calculated by the Forestry Agency.
Source: Forestry Agency (1994), *Heisei 5 Nendo Ringyo Hakusho*, 192.

Akao (1993) estimated cost functions for planting, silviculture, harvesting, and transportation, respectively. He also constructed a "business as usual" scenario for future wage rates in forestry as well as stumpage prices by species and stand age. Integrating these factors, he calculated optimal cutting age and land value based on the deterministic Faustmann formula for all 248 forest units of sugi and 396 forest units of hinoki, where each "forest unit" denotes an area covered with trees that are the same in terms of species and age. This calculation took account of the characteristics of each forest unit: fertility index and distance from the nearest forest road. A discount rate of 3.5% was used, the standard rate at which the Agriculture, Forestry and Fisheries Finance Corporation provides funds.

It was found that for about a quarter of the forest units, the optimal strategy is to leave the stand alone and not continually cut it. For the remaining forest units, there is an optimal cutting age – for sugi, around 45 years (30% of the forests) and 80 years (70%); for hinoki, around 54 years – but it is not optimal to plant again after harvesting the current stock. These results may be rather pessimistic, since the calculation did not take into account subsidies and price uncertainty surrounding forestry. For example, if uncertainty were introduced in an appropriate way, expected revenues from harvesting would be greater, since one can have a chance in stochastic fluctuation, possibly more than twice that calculated by Akao (Brazee and Mendelsohn 1988). On the other hand, it must be pointed out that the results for regeneration scarcely varied, even if 2.5% was used as the discount rate.

The statistical results show that the implications of the severity of market conditions relating to private forestry go far beyond the ability to adjust by selecting cutting age. They also suggest that most forest managers may leave forestry and abandon their forestland once the current stands have been felled. Private forestry in Japan may vanish almost entirely with the harvesting of trees planted during active reforestation in the postwar period.

Such is the impact of the market system on an industry where technological innovation has not occurred and/or that has lost its comparative advantage in international competition. Forests, however, have different environmental functions, such as flood control, scenic beauty, genetic pool, and so on. If most forestland is not planted after current stands are felled, society will suffer significantly from the loss of the environmental benefits provided by forests.

In the context of global environmental issues, another important task is to realize sustainable forest management. To abandon land would be contrary to international objectives. The government will therefore be under pressure to conserve private forests more than before. This will be in the government's best interests, since the environment is now one of the most important concerns of the average citizen. As a consequence, more subsidies may be provided to private forests, and public sector or non-governmental organizations may begin to operate forests for environmental purposes, leasing forestland from owners in order to do so. Thus privately owned forests will survive, but private forest management making a profit from timber harvesting may become virtually extinct.

References

Akao, Ken-ichi (1993). "Bunshu zorin keiyaku to shakaiteki saiteki bakkirei – aru shinrin seibi houjin no jirei bunseki –" ("Profit-sharing reforestation contract and social optimal cutting ages. A case study of a local public forest corporation"), *Kyoto Daigaku Enshurin Houkoku (Bulletin of the Kyoto University Forests)* (65): 194-209 (in Japanese with English summary).

– (1996). "Stochastic forest management and risk aversion," *Journal of Forest Planning* 2: 131-36.

Brazee, Robert, and Mendelsohn, Robert (1988). "Timber harvesting with fluctuating prices," *Forest Science* 34: 359-72.

Johansson, Per-Olov, and Löfgren, Karl-Gustaf (1985). *The Economics of Forestry and Natural Resources.* 292 pp. Basil Blackwell, Oxford.

Minamikata, Yasushi (1995). "Kouseino ringyo kikaika 7 nen wo keika shite" ("An analysis about structure of lumber products"), *Ringyo Keizai (Forest Economy)* (565): 10-20 (in Japanese).

Sato, Nobuko (1999). "Rinka keiei, tokusan ron." In Funakoshi, Shoji, ed., *Shinrin, Ringyo, Sanson Mondai Kenkyu Nyumon,* 120-34. Chikyusha, Tokyo (in Japanese).

4
Forest Owners' Associations
Koji Matsushita and Kunihiro Hirata

Shinrin Kumiai, forest owners' associations whose members are private forest owners, were first established under Japan's 1907 Forest Law (also called the Second Forest Law). The direct English translation is "forest cooperative," but since the *shinrin kumiai* are thought of as associations of forest owners, the term "forest owners' association" is used in this chapter. (Several chapters in this book translate *shinrin kumiai* as "forest cooperative" and some papers use both "forest owners' association" and "forest cooperative.")

Considering the current state of forestry and forest resources, forest owners' associations occupy an important place in Japan. Three examples illustrate this:

- At the end of fiscal 1994, 57.9% of the total forest area in Japan was private forest, belonging to many types of owners, including individuals, companies, shrines, and temples. Numerically, the largest group of private forest owners consists of individuals; there were 2,508,000 in the 1990 Agriculture and Forestry Census. Forest owners are classified by the area they hold: 57.9% own less than 1 ha of forest and 31.0% own 1-5 ha. Approximately 90% own forests of less than 5 ha, so there are clearly many small-scale owners of private forest. Forest owners' associations include a large number of these small-scale private forest owners.
- The population of Japan's mountainous regions has been decreasing, while the average age of forest owners is increasing. Under these conditions, it is gradually becoming more difficult for private forest owners to manage their forests, and the forest owners' associations are therefore expected to manage private forests, instead of the owners. The Forestry Agency has spent large amounts to subsidize these associations.
- The depopulation of the mountainous regions has resulted in a decrease in the number of forestry workers even as the age of the remaining workers is increasing. The role of forestry workers employed by forest owners' associations is increasing.

Thus, forest owners' associations have become very important, both as organizations of private forest owners and as the lowest level of forestry administration.

This chapter has three parts. First, the institutional history of forest owners' associations since the passage of the 1907 Forest Law is presented. The second part outlines forest owners' associations and their volume of business, including the forestry working units. Finally, the institutional problems related to forest owners' associations are discussed, especially: (1) the relationship between forest owners' associations and forestry administrative organizations, (2) forest management by the forest owners' associations, and (3) the relationship between the forest owners' associations and private forest owners.

Historical Background

Forest Owners' Associations before the Second World War

1907 Forest Law
Forest owners' associations were first established under the 1907 Forest Law (Law No. 43, also called the Second Forest Law). Before 1907, local legislation had established organizations of forest owners that were substantially the same as the forest owners' associations defined by the 1907 Forest Law. The first association was established in Toyama Prefecture in 1885. The aims of the associations established before the 1907 legislation varied. The association established in Hyogo Prefecture in 1891 prevented burning and destructive cutting. The association established in Hiroshima Prefecture in 1894 controlled tree cutting, the digging up of soil and stone, and forestland reclamation; there was also an agreement on harvesting and utilizing undergrowth and leaf fall (Kaibara 1955). Because the area of denuded forestland had spread after the Meiji restoration, most of the owners' associations established at that time aimed to protect forest resources. The 1896 government bill that was to become what is called the First Forest Law included provisions for forest owners' associations. These were ultimately deleted when the First Forest Law was enacted in 1897 (Law No. 29), because of opposition to the planned compulsory establishment of, and participation in, forest owners' associations, and because national and public forests were excluded from membership in these associations. Provisions for forest owners' associations were prescribed in the Second Forest Law instead.

The main purpose of the 1897 Forest Law was to regulate forestry practice, and its main provision was the introduction of a forest protection system, called *hoanrin* in Japanese. The Forest Law was amended in 1907 to promote forestry that would meet the increasing demand for wood. This new law included the first provisions for forest owners' associations, as organizations of forest owners that worked on forest practices jointly. It provided for four types of such associations: forest silviculture, practice, road construction, and protection.

An association for silviculture could be established when joint reforestation was necessary for the recovery of denuded forestland. An association for forest practice could be established when necessary for joint management plans for forests with multiple owners. An association for forest road construction could be established for the construction and maintenance of forest roads and related facilities. An association for forest protection could be established when joint work was necessary to prevent fire, illegal cutting, insect infestations, and so on. These four categories could be combined freely.

Two conditions had to be met to establish a forest owners' association. The agreement of at least two-thirds of the total number of forest owners within the jurisdiction of the association was required, and the total forest area owned by these owners had to exceed two-thirds of the total forest area within the jurisdiction of the association. In this sense, establishment of an association was discretionary. When a forest owners' association was established, however, all the forest owners within its jurisdiction had to join it. At the time the 1907 Forest Law was adopted, few forest owners' associations had been established. By 1939, when it was amended, there were 2,654 such associations based on the 1907 law. Most of those established by the 1907 law had been established by local governments, and great importance was attached to public benefits rather than the pursuit of private benefits (Shioya 1973). The most common associations were for forest practice and forest-road construction; approximately 1,500 of each type were established. (Since forest owners' associations could involve multiple categories, the sum of the totals for each type exceeds the total number of forest owners' associations.) Most associations for forest practice did not have a forest management plan, and most associations for forest-road construction became almost nominal entities after the forest road was constructed.

1939 Amendment of the Forest Law

The Forest Law was amended in 1939 (Law No. 15) to comply with the National Mobilization Law of 1938. The 1939 Amendment, as it is called, required forest management plans for all non-national forests, and the forest owners' associations were to have an important role in drawing them up. An owner who managed over 50 ha could make an individual forest management plan; otherwise, a forest owners' association had to make the plan.

Forest owners' associations were reclassified into two categories. In the first, the association made the forest management plan for the forest that the association members owned, and managed forest practices directly. In the second category, the association made the forest management plan for the forest owned by its members, but only instructed members about the forest practices required.

Because it was necessary to make a forest management plan for all non-national forests, a new system of compulsory establishment and compulsory participation was introduced. The establishment unit became broader, and was municipality-based in principle. Furthermore, the associations began to develop relationships with administrative units. The 1939 Amendment increased the number of forest owners' associations, which reached 5,822 in 1951, when the Forest Law was next amended. In order to control the associations established on the basis of municipalities, prefectural and national forest owners' associations were established. With the enforcement of the Timber Control Law in 1941 (Law No. 66), the associations became the lowest level of timber control administration.

Forest Owners' Associations after the Second World War

1951 Forest Law

The 1951 Forest Law is also called the Third Forest Law. Its main objective was to promote the recovery of forest resources that had been depleted by destructive cutting during and after the war. The provisions for forest owners' associations were amended to democratize the organizations. Under the direction of GHQ (General Headquarters of the Allied Powers), forest owners' associations had to be based on cooperative principles, as laid out in a 1950 statement (statement of Fair Trade Practices Division, Economic and Scientific Section, and Forestry Division, Natural Resources Section, presented to officials of the Forestry Agency, Ministry of Agriculture and Forestry, on 25 February 1950). Three main points in this statement called for the government to:

- "1. Revise the Forest Law of 1907 to place responsibility for formulation and implementation of forestry programs more definitely in the central government, with the provision that its authority may be delegated to the prefectural and local governments"
- "6. Delete from the Forest Law the authority of the forest owners' associations to engage in business activities," and
- "7. Provide legislation to enable groups of individual forest owners to carry on their business activities in concert on the basis of cooperative principles and for the establishment of federations of such cooperatives." (Zenkoku Shinrin Kumiai Rengokai 1973)

The establishment of forest owners' associations became discretionary, and forest owners could join and leave the associations freely. Each forest owner had the right to one vote and could invest freely in more than one unit. Point 1 of the statement abolished the system by which forest owners' associations made forest management plans, introduced in the 1939 Amendment. Forest owners' associations were classified into two categories: *shinrin kumiai,* which cooperated in managing part of a forest, and *seisan shinrin kumiai,* which managed forests as a forest owner.

The 1951 Forest Law prescribed that forest owners' associations had an obligation to conduct at least one business among three business categories: Guidance, Forest Management, and Trusted Forest Management (these categories will be described later). In addition, the business of an association was strictly limited to being directly related to forestry in general. In contrast, there were no indispensable business restrictions on agricultural cooperative associations called *nokyo,* which were also established as cooperative organizations. Furthermore, agricultural cooperative associations were allowed to deal with deposits and savings, whereas forest owners' associations were not. There are, then, clear differences between forest owners' associations and agricultural cooperative associations.

The business of forest owners' associations was restricted because of the nature of forest owners' associations as defined in the 1951 Forest Law. Article 74 of this law laid out the two main aims of forest owners' associations: (1) to promote the rationalization of forest practices and increase forest productivity, and (2) to improve forest owners' socio-economic position. Point 1 is related to public benefits, and point 2 to private benefits. Accordingly, the forest owners' associations prescribed by the 1951 Forest Law were designed to be cooperatives involving individual forest owners, although the first principle of the association involved public benefit. The forest owners' associations established by the 1939 Amendment could become associations as prescribed by the 1951 Forest Law by changing their articles of association. There were 5,029 forest owners' associations in 1952. Before 1952, all the associations were *shinrin kumiai;* those classified as *seisan shinrin kumiai* were established after fiscal 1953. After the Second World War, forest owners' associations were not actively involved in business in general, and their role was limited to complementing forestry administration (Tanaka 1996).

The Basic Forestry Law (Law No. 161) was enacted in 1964. Its purpose was to increase total forestry production, forest productivity, and the incomes of forestry workers (Article 2). Before this law was enacted, legislation relating to forestry and forest resources was based on the Forest Law, which was mainly concerned with sustaining the yield of forest products and realizing non-timber values. Although the Forest Law also dealt with economic aspects, including the articles related to forest owners' associations, these were of secondary importance. With the boom in the Japanese economy, and growth in the fields of agriculture, forestry, and fisheries, legislation for the purpose of economic development progressed, and it was in this environment that the Basic Forestry Law was enacted. Since the private forestry sector in Japan is characterized by small average forest holdings, Article 13 of the Basic Forestry Law provided for the promotion of cooperation in forest practices and forest management by forest owners' associations. It was important that the role of these associations was clearly stated in law.

To comply with the Basic Forestry Law, the Forest Law was amended in 1974 and the purpose of forest owners' associations was changed (Article 74). Its primary purposes were: (1) to improve forest owners' socio-economic position, and (2) to promote the rationalization of forest practices and increase forest productivity. The 1974 Amendment expanded the types of business that forest owners' associations could provide. For example, associations could now sell garden stones, crushed stone, and plants for reforestation. New areas of business included forestland supply, facilities for forest product processing, and forest recreation.

Enacting the Forest Owners' Association Law
The provisions for forest owners' associations were removed from the Forest Law in 1978, and a new law, the Forest Owners' Association Law (Law No. 36), was enacted the same year.

After the 1974 Amendment, including the change in the legal position of forest owners' associations, the Forestry Agency organized a committee to look into the system of forest owners' associations and began to review their organization and roles. It was argued that the associations needed to be strengthened to play an important role in local forestry development (Shinrin Kumiai Seido Kenkyukai 1998). Article 1 of the Forest Owners' Association Law stated two equally important purposes for such associations: (1) to increase the socio-economic position of forest owners, and (2) to sustain forest yields and increase forest productivity. Point 1 was based on the Basic Forestry Law and point 2 on the Forest Law, meaning that the political position of forest owners' associations was based on both forest resource policy and forestry economic policy. The Forest Owners' Association Law of 1978 made it clear that these associations are both public and cooperative organizations (Shinrin Kumiai Seido Kenkyukai 1998).

The kinds of forest owners' associations were also changed in the 1978 legislation. Under the 1951 Forest Law, there were two categories, one of which was *seisan shinrin kumiai*. The new legislation removed *seisan shinrin kumiai* from the forest owners' association system. Most of the *seisan shinrin kumiai* were established in order to clarify the proprietary rights of common forests, called *iriairin*, in which residents used forest resources, including all kinds of forest products, according to local custom. In this sense, *seisan shinrin kumiai* were very different from the usual forest owners' associations, *shinrin kumiai*. The two systems were therefore separated as a matter of forest and forestry policy. We will not refer to *seisan shinrin kumiai* again in this chapter. In the 1978 legislation, forest protection was added as one of the business categories from which forest owners' associations were obliged to select at least one business. Forest management and practices were also promoted.

The current Forest Owners' Association Law has been amended several times since 1978, mainly in 1987 and 1997. The major amendments in 1987 concerned three points (Shinrin Kumiai-ho Kenkyukai 1989):

- *Expansion of the areas of business.* These included the negotiation of loans to members of the association, the purchase of goods for forest owners, the construction of buildings using timber produced by forest owners, the production of edible mushrooms, the promotion of forest management trusts through the prefectural forest owners' associations, and increasing the capacity of the suretyship obligation of the prefectural forest owners' association.
- *Introduction of the Joint Forest Practice Regulation Program, prescribed by Article 25(2) of the Forest Owners' Association Law.* This regulation, which covers joint work on forest practices based on contracts among forest owners, was made for forests that must be improved jointly in order to achieve a sustained yield or increase forest productivity. In effect, the forest owners' association clarifies basic management policies in order to promote joint work on forest practices in non-national forests by forest owners. When a forest owner participating in the agreement requires forest practice according to the program, the association has to conduct forest practice according to the agreement. The association can encourage owners of forest adjoining an area covered by such an agreement to join the program.
- *Improvement of the forest owners' association management system.*

The major amendments in 1997 involved three points (Forestry Agency 1997b):

- *Expansion of the areas of business.* Before the amendment, the processing and sales departments of the association were limited to dealing with forest products; this was expanded to include general commodities such as agricultural products. Previously, facilities jointly used by forest owners were limited to those related to forestry; this was expanded to include general facilities necessary for the business and life of forest owners, such as facilities for an agricultural contracting business involving labourers from the forest owners' association.
- *Introduction of specified forest owners' associations.* For any Regional Forest Planning Area in which forest improvement projects are underdeveloped,

the prefectural government can specify a forest owners' association, which can then implement forest practices efficiently for the whole planning area. This new policy mainly targeted 706 municipalities, located near urban areas, that had no forest owners' associations within their administrative jurisdictions.

- *Improvement of the organization of the forest owners' associations.* The existing organizational system was implemented before the Second World War. The administrative unit when the forest owners' association system was legislatively introduced was the village. Recently, the jurisdiction has expanded to municipalities or multiple municipalities with amalgamation. In addition, the types of business in which associations are involved have diversified. These improvements were necessary in order to increase efficiency and keep step with changing times.

Promotion of the Amalgamation of Forest Owners' Associations
In the early 1950s, when the system of forest owners' associations was introduced based on the 1951 Forest Law, there were over 5,000 associations. A significant number of these conducted practically no business. In 1963 the Forest Owners' Association Amalgamation Promotion Law (Law No. 56) was promulgated to encourage the amalgamation of these associations, and the Forestry Agency subsequently promoted this amalgamation. Since the passage of this legislation, the number of forest owners' associations has steadily declined: 3,417 in fiscal 1963; 3,077 in fiscal 1965; 2,524 in fiscal 1970; 2,187 in fiscal 1975; 1,933 in fiscal 1980; 1,790 in fiscal 1985; 1,642 in fiscal 1990; 1,455 in fiscal 1995; and 1,254 in fiscal 1999. The number of associations with jurisdiction in more than one municipality has increased, but remains less than one-quarter of all the associations.

The Forestry Agency (1997a) set a goal of 600 forest owners' associations by the end of fiscal 2001, but achieving this goal is now impossible. The agency gave two reasons why the amalgamation of associations has not progressed according to the timetable. The first is the financial disparity between forest owners' associations. It is generally difficult to amalgamate an actively managed forest owners' association with one that does almost no business. The other reason is related to manpower, and is a more difficult problem. The management staff and membership of a forest owners' association are often unaware of difficulties in surrounding associations. Furthermore, there is rarely a strong leader capable of bringing about the amalgamation of the associations; the position of director of a forest owners' association is an honorary post, and frequently the municipal leader or a member of the municipal assembly fills this position. Under such circumstances, an amalgamated forest owners' association often has branch offices in every area of the pre-amalgamation associations, meaning that the amalgamation has produced virtually no change. Consequently, the amalgamation of associations with conflicting management styles and levels of business sometimes results in an ineffective association (Fujita 1998).

After the 1991 amendment of the Forest Law, which introduced a new basic forest policy based on river basins, the jurisdiction of amalgamated forest owners' associations was expected to match that of Regional Forest Planning Areas, which were reorganized based on the major river basins. The Forestry Agency (1997a) pointed out the possibilities of a wider area of amalgamation, such as having one forest owners' association for some prefectures.

Business and Forestry Labour

Business Categories

Types of Business
The types of business that forest owners' associations can practise are set out in the Forest Owners' Association Law (mainly Article 9). The current business categories include: (1) Guidance, (2) Sales, (3) Forest Production, (4) Forest Products Processing, (5) Revegetation Plants, (6) Construction, (7) Co-op Business, (8) Nursery Stock, (9) Reforestation, (10) Forest Protection, Forest Recreation, and Miscellaneous, (11) Forestland Dealings, (12) Welfare, (13) Finance, (14) Forestland Disposition by Sale, (15) Forest Management, and (16) Trusted Forest Management. Each category is described briefly next (Forestry Agency 1999a).

Guidance includes activities such as guiding forest management, educating forest owners, disseminating information, and encouraging owners to join the Joint Forest Practice Regulation Program. The most important role of guidance in forest management is to facilitate a Forest Operation Plan, as prescribed in the Forest Law. Information dissemination is mainly concerned with subsidies for forests and forestry.

Sales involves the sale of forest products and revegetation plants. The sale of logs cut by forest owners or logging contractors is also entrusted to forest owners' associations.

Forest Production includes the production and sale of forest products. It is different from Sales in that it includes everything from tree cutting to log sales.

Forest Products Processing deals with the processing and sale of forest products, specifically the management of sawmills and factories processing forest by-products.

Revegetation Plants include activities such as harvesting, growing, processing, and sales.

Construction relates to the construction and sale of buildings, such as housing.

Co-op Business is the cooperative buying of materials necessary for forestry or other business as well as for the daily life of forest owners. This might include forestry tools, such as hatchets, sickles, and chainsaws, and planting stock.

Nursery Stock includes the production and sale of seeds and planting stock.

Reforestation includes reforestation, the setting up of forest management trusts, erosion control projects, and forest road construction. The last two items are strictly connected to public works. Forest management trusts are limited to public forests. These trusts deal with reforestation and forest management after reforestation.

Forest Protection, Forest Recreation, and Miscellaneous include the activities of disease and insect control, such as the cutting of damaged trees and chemical spraying, storage and the transportation of forest products, forest surveys, office work involved in applying for reforestation subsidies, recreational facilities related to forests, and the making of Forest Operation Plans.

Forestland Dealings are the activities to mediate dealings of forestland at the request of members.

Welfare is divided into two areas of concern: the promotion of safety and health in forestry work, and forest insurance.

Finance involves loans to forest owners that are necessary for forestry or other businesses or for daily life. Forest owners' associations can borrow money from the Agriculture, Forestry and Fisheries Finance Corporation, and lend it to forest owners.

Forestland Disposition by Sale deals with the sale of forestland for other uses.

Both *Forest Management* and *Trusted Forest Management* involve forest management by the forest owners' association.

As outlined above, the types of business that forest owners' associations are involved in have expanded greatly. In fiscal 1999, total transactions involved ¥353.8 billion. Of this, Reforestation generated ¥179.9 billion, Sales and Forest Production ¥83.7 billion, Forest Products Processing ¥40.1 billion, Co-op Business ¥17.6 billion, and all other areas ¥32.5 billion. Financially, the most important business categories are Reforestation, Sales, and Forest Production. These businesses include tree cutting and log sales by forest owners, and planting, erosion control projects, and forest road construction, which are supported by government subsidies and public works.

The volumes of the main business categories related to forestry, excluding erosion control and forest road construction, are shown in Table 4.1. The number of businesses involved in Sales and Forest Production, which are the main businesses of forest owners' associations, has declined, especially in the case of Forest Production. The volume in fiscal 1999 was down 29.4% from the peak volume in fiscal 1987. Looking at the proportion of transactions by forest owners' associations in relation to the total dealings for private forests throughout Japan, the share related to reforestation is

Table 4.1

Main business categories of forest owners' associations in Japan, 1952-99.

Fiscal year	Sales Log[1]	Forest production Log[1]	Forest products processing Lumber[1]	Nursery stock Planting stock[2]	Reforestation Planting[3]	Tending[3]
1952	132	360	1,227	444	69.2	13
1953	615	790	299	506	43.3	23
1954	156	279	276	523	–	–
1955[4]	–	–	–	–	–	–
1956	233	292	219	525	10.2	10
1957	330	324	222	568	9.1	13
1958	431	376	222	662	11.5	19
1959	538	532	279	743	17.3	27
1960	618	449	201	795	13.5	31
1961	698	716	198	856	19.9	58
1962	715	822	166	873	24.0	69
1963	703	1,058	183	914	31.5	83
1964	786	1,247	164	799	30.2	118
1965	849	1,342	150	757	39.6	137
1966	925	1,512	154	754	43.3	164
1967	960	1,509	156	824	52.7	185
1968	869	1,597	160	806	61.6	236
1969	1,034	1,869	148	803	67.8	277
1970	1,008	2,042	178	730	71.6	306
1971	1,053	2,260	111	628	76.4	350
1972	1,053	2,220	187	548	77.9	389
1973	927	2,243	209	515	75.6	427
1974	941	2,012	176	476	75.9	457
1975	858	2,090	201	446	72.1	498
1976	1,005	2,177	206	415	67.4	536
1977	1,140	2,285	210	390	68.3	566
1978	1,112	2,225	227	362	67.2	605
1979	1,186	2,275	232	337	71.8	691
1980	1,291	2,179	177	326	75.8	776
1981	1,286	2,211	182	298	74.6	834
1982	1,521	2,343	205	285	76.5	817
1983	1,453	2,379	204	257	73.7	806
1984	1,550	2,963	273	238	67.2	833
1985	1,702	3,576	285	214	61.7	812
1986	1,658	3,799	314	187	57.7	816
1987	1,728	3,981	354	174	53.2	804
1988	1,702	3,826	397	173	50.6	789
1989	1,749	3,754	385	151	48.7	765

▶

◄   *Table 4.1*

Fiscal year	Sales Log[1]	Forest production Log[1]	Forest products processing Lumber[1]	Nursery stock Planting stock[2]	Reforestation Planting[3]	Tending[3]
1990	1,870	3,282	407	140	45.4	766
1991	1,823	3,194	423	129	40.9	740
1992	1,863	3,350	448	124	39.1	714
1993	1,827	3,293	504	120	37.5	692
1994	1,868	3,336	541	204	36.3	651
1995	1,799	3,088	515	116	35.2	629
1996	2,017	3,172	537	105	31.8	602
1997	1,962	3,110	521	100	31.0	569
1998	1,716	2,778	492	171	29.9	566
1999	1,919	2,810	513	197	26.5	547

1 In thousands of cubic metres.
2 In millions.
3 In thousands of hectares.
4 No data.
Sources: Forestry Agency, 1952-2001, *Statistics of Forest Owners' Association* (*Shinrin kumiai tokei*);
Forestry Agency, 1964-2001, *Statistical Handbook of Forestry* (*Ringyo tokei yoran*); Statistics and
Information Department, Ministry of Agriculture, Forestry, and Fisheries (1983), *Portable Size
Statistics of Forestry* (*Pocket ringyo tokei, ruinenban*).

high, approximately 90% of planting and 80% of thinning (Forestry Agency
1997b). Recently, the associations' share in log production was approximately 15% (Forestry Agency 1999b). The share in forest products processing is still low in spite of the recent increase in volume.

Expansion of Business Areas
When the forest owners' association system started after the Second World
War, the types of business that they could be involved in were strictly limited. The types of business were expanded in the 1974 Amendment to the
Forest Law and in the 1987 and 1997 amendments to the Forest Owners'
Association Law. The scope of business has been expanded legislatively because future growth is not expected in the volume of business directly related to forestry, due mainly to long-term sluggish trends in the price of
forest products. The volume involved in reforestation, the largest business
category, will decrease in the future because of decreases in the areas of
afforestation, reforestation, and clear-cutting, and a slowdown in public
works related to forestry.

Forest owners' associations have long been expected to deal with deposits
and savings. It was argued that this should be included in the 1978 Forest

Owners' Association Law and the 1987 Amendment, but it was deemed too early to legislate this. Concerning forest insurance, there were two systems before 1995: Government Forest Insurance (started in 1937), managed by the Forestry Agency, and Mutual Relief of Forest Damage (started in 1956), managed by the National Federation of Forest Owners' Associations. Since both insurance systems had mutual concerns and forest owners' associations did the office work for both systems, the Administrative Management Agency advised that they be unified in 1972 (Matsushita 1996a). In 1995, a new forest insurance system was started that maintains both systems, as each system now provides 50% of the insurance coverage.

With the amendment of the Forest Owners' Association Law in 1997, forest owners' associations recently started new types of business. The Forestry Agency (1998) offered four examples:

- In Wakayama Prefecture, a forest owners' association began harvesting the Japanese apricot, which is a special product of the village located in this forest owners' association. Because of depopulation, there was a shortage of agricultural labour, whereas it was difficult for the association to employ forestry workers year-round in work related only to forestry.
- In Nagano Prefecture, the forest owners' association began dealing in landscape gardening, which is a business related to trees.
- In Niigata Prefecture, a forest owners' association started doing snow removal work in order to employ forest owners' association workers in winter.
- Finally, in Gifu Prefecture, the forest owners' association began harvesting the cogon grasses used to thatch the roofs of traditional buildings in Shirakawa District. These buildings were registered as conservation sites under the protection of the World Cultural and Nature Heritage in 1995.

None of these businesses were directly connected to forestry, but in each case the forest owners' association started a new type of business to provide more employment.

Forestry Labour

Forestry Workers

Forest owners' associations employ two categories of worker for forestry practices such as cutting and planting: employees of the forest owners' association or contract workers from outside the association. Currently, approximately 80% of associations employ their own forestry workers. Some of these are employed by the association full-time, while others work part-time to supplement their own farm or forestry incomes.

According to the Forest Owners' Association Statistics for fiscal 1999, there were 1,254 associations and 1,222 responded to a questionnaire. Of those responding, 996 (81.5%) employ their own forestry workers. There are a total of 30,680 forestry workers, employed for various periods each year: 4,732 (15.4%) for less than 60 days annually; 2,024 (6.6%) for 60-89 days; 4,992 (16.3%) for 90-149 days; 7,715 (25.1%) for 150-209 days; and 11,217 (36.6%) for 210 or more days. Numbers of forestry workers by age class are as follows: 1,808 (5.9%) less than 30 years old; 2,222 (7.2%) 30-39 years old; 4,097 (13.4%) 40-49 years old; 6,951 (22.7%) 50-59 years old; and 15,602 (50.8%) 60 or more years old. Note that more than 50% of forestry workers were over 60 years old, while very few were less than 30 years old.

At one time, forest owners' associations did not directly employ forestry workers and all work was done by outside workers. Beginning in the mid-1960s, associations started to employ their own workers. The number of workers employed directly by associations increased steadily until it peaked in 1970: 15,800 in fiscal 1963; 32,200 in fiscal 1964; 38,200 in fiscal 1965; 44,000 in fiscal 1966; 53,000 in fiscal 1967; 63,300 in fiscal 1968; and 67,100 in fiscal 1969 and 1970 (Forest Owners' Association Statistics). This increase was due to the passage in 1964 of the Basic Forestry Law, which introduced the Forestry Structure Improvement Project, as a result of which more associations conducted forest practices, such as planting and tending. There were approximately 60,000 forestry workers during the 1970s and early 1980s. The number peaked in fiscal 1982, and declined to 30,680 workers by fiscal 1999, mainly because of the aging of forestry workers.

Forest owners' association forestry workers form the largest group of forestry workers in Japan. According to the Agriculture and Forestry Census of 1990, 76,900 employees were engaged in forestry works for over 150 days a year. Of these, 27,200 (35.4%) were forest owners' association forestry workers. The other large categories were employees of companies (18,900 persons, or 24.6%) and the district forestry offices of the Forestry Agency (13,300 persons, or 17.3%). Since the forest owners' associations play an important role in planting and tending in non-national forests, national and prefectural governments have promoted several programs to encourage them to hire forestry workers. Nevertheless, the aging trend and decrease in numbers continue.

Employment Conditions
The employment conditions of forestry workers employed by forest owners' associations were learned through a survey conducted in the Hokusatsu Forest Planning Area of Kagoshima Prefecture in fiscal 1995 (Matsushita 1995, 1996b). The survey was given to 309 forestry workers employed by

forest owners' associations. There were 222 responses, 136 from men and 84 from women. The workers' average ages were 58.3 years for men and 59.0 years for women. The proportion of forestry workers over age 50 was 86.0% for men and 89.3% for women. The proportions for those over age 60 were 60.3% and 56.0%, respectively. The average duration of service was 13.7 years for men and 15.3 years for women. The proportion of forestry workers employed for less than 10 years was 26.6% and their average age was 55.6 years, which is only slightly less than the average age of all forestry workers. This indicates that the average age of newly appointed workers is also high. Looking into the future, 55.9% indicated that they would retire within 10 years, and 45.0% pointed out that the greater age of workers results in a decrease in work efficiency.

The forestry workers indicated that the main problems with their employment conditions were wages (31.5%), social security (8.6%), employment security (8.6%), and others (3.6%). The greatest dissatisfaction was with wages. Most forestry workers were paid a daily/monthly wage (68.5%), while some were paid by piecework (13.1%). Under the daily/monthly wage system, the total wage is calculated by multiplying the daily wage by the number of days worked in a month. The average daily wage was ¥7,174 for men and ¥5,867 for women. The rate differed with the kind of work. For example, men doing cutting were paid an average wage of ¥7,825. These forestry workers would have liked to be paid an average of ¥11,193, so the difference between the desired and actual wage was very large. The estimated annual income from forest owners' associations, calculated by multiplying the average daily wage by the annual number of working days, was ¥1,400,000 for men and ¥1,080,000 for women. The estimated average annual household income, calculated from the estimated annual income from forest owners' associations and the share of that income in the total household income, was ¥2,570,000 for men and ¥2,270,000 for women. These are very low incomes in Japan; in 1998, the lowest classification by yearly income quintile group in the Family Income and Expenditure Survey of the Statistics Bureau, Management and Coordination Agency, was an income below ¥4,050,000.

By law, all workers must be on the workers' accident compensation insurance list. Of the other social security systems, the percentages of forestry workers covered are as follows: National Pension, 65.6%; Agriculture and Forestry Pension, 8.6%; Welfare Pension Insurance, 4.5%; National Health Insurance Association, 6.6%; other health insurance, 18.6%; Forestry Retirement Allowance Mutual Aid Association, 51.6%; and Employment Insurance, 52.9%. Most forestry workers are inadequately covered by most of the basic social security systems, such as pension, health insurance, retirement allowance, and employment insurance.

In spite of low wages and inadequate social security, most forestry workers continue to work for forest owners' associations because they are too old to work elsewhere and they can supplement their income by farming (58.4% of forestry workers obtain some income from agriculture). According to the questionnaire, most forestry workers do not want to work more days annually, nor do they want to pay increased premiums for social security. With such poor wages and social security, however, it is very difficult to attract new labour, especially young labour.

Characteristics and Problems of Forest Owners' Associations

Forest Owners' Associations and Administrative Organization

The 1939 Amendment to the Forest Law included the forest owners' association system with war organizations. The most important change in 1939 was that the unit of establishment for forest owners' associations became municipalities; the relationship between forest owners' associations and municipal offices thus became clear. According to the 1939 Amendment, the associations were to play an important role in making forest management plans for non-national forests on behalf of many small-scale forest owners. The association became the lowest-level administrative unit for non-national forests, and acted in conjunction with municipal forestry departments. After the Second World War, forest owners' associations were reorganized by introducing cooperative principles under the direction of GHQ. They remained, however, the lowest-level unit for forestry administration.

This section discusses (1) the relationship between forest owners' associations and the forest planning system, which is the most important forest resource policy, and (2) the relationship between forest owners' associations and municipal forestry departments.

Forest Owners' Associations and the Forest Planning System
Although the prewar governmental system of supervising forest management planning changed after the war, the importance of forest owners' associations remained unchanged to the present. The current forest planning system for non-national forests consists of the National Forest Plan (formulated by the Forestry Agency); Regional Forest Plans (formulated by the prefectural government), for which the forest planning area is based on river basins; and Forest Operation Plans (formulated by forest owners). There are two types of Forest Operation Plans: Forest Operation Plans made by individual forest owners and Territorial Joint Forest Operation Plans, made by multiple forest owners for all the forests in a given area. In effect, the latter are formulated by forest owners' associations.

Since the forest planning system under the current Forest Law is basically a planning system for forest resources, it is important that forest owners' associations formulate Forest Operation Plans in areas where individual small-scale forest owners cannot formulate their own. This situation is quite different from that before the Second World War, which was directly connected to the wartime control system. It is worth noting that the National Forest Plan and Regional Forest Plan are strictly related to subsidies related to forest policy, such as planting, tending, and forest road construction. Since the Forest Operation Plan is the lowest level of the forest planning system under the Forest Law, it is connected to the entire policy program, including subsidies from the national and prefectural governments. A forest owners' association makes a Forest Operation Plan on behalf of its members, and simultaneously applies for subsidies for forest practices.

This system, in which forest owners' associations mediate between the prefectural departments of forestry and forest owners, appears to have two advantages for forestry administration: (1) the Forest Operation Plan can include a larger area of private forest, and (2) it becomes possible to administer forest resources and forestry based on the forest planning system, especially the allocation of subsidies. This system also has problems, however. The strong administrative systematization of the prefectural government, forest owners' association, and forest owners means that the associations have a virtual monopoly on obtaining subsidies related to effective forest practices, and the management of forest owners' associations is dependent on the subsidies. It appears to be extremely difficult for the associations to develop as enterprises when they continue to depend on subsidies and function within the limits of these subsidies. If a repeated annual volume of business is secured by subsidies from the prefectural government, obtained without special effort, then the forest owners' associations have little incentive to increase business, unlike most typical companies.

It is possible that this situation also influences the business activities of forestry-related enterprises other than forest owners' associations. For example, a forest owners' association can easily obtain subsidies to improve log production and sawmill facilities. The most important subsidy is from the Forestry Structure Improvement Project based on the Basic Forestry Law. Consequently, forest owners' associations have an advantage over private enterprises such as logging contractors and sawmill companies. There appears to be no clear justification for giving these associations such an advantage in forestry enterprises.

Forest Owners' Associations and Municipal Departments of Forestry
Since the postwar forest owners' association system was based on the system set out in the 1939 Amendment to the Forest Law, for which the establishment unit was the municipality, the postwar associations were still the lowest level of forestry administration. Unfortunately, the relationship between them and municipalities is not clear.

When the 1951 Forest Law was enacted, the municipal department of forestry played a role in establishing clear national-level objectives. These objectives were to restore national land that had suffered from heavy cutting during and immediately after the war, and to increase the future production capacity of coniferous forests. Because the economic importance of agriculture and forestry has gradually decreased, however, forestry-related administration has occupied a decreasing proportion of the total municipal administration. The number of municipalities that have a specialized forestry department, such as a department of forestry or a department of forest policy, has been decreasing. In Kagoshima Prefecture, only one of 96 municipalities has a department of forest policy (Matsushita 1999). In most cases, the Department of Agriculture, Forestry, and Fisheries or the Department of Industrial Development is also responsible for the forestry policy program.

The number of full-time forestry staff has also been decreasing. In Kagoshima Prefecture, municipalities have an average of 0.93 full-time forestry staff. Only 30% of municipalities have at least one staff member who graduated from forestry high school or a university department of forestry, and most municipal offices have no staff with a forestry education. One reason that such organizational weakening in the municipal office was possible was that forest owners' associations played an important role in real forestry administration, by handling applications for subsidies related to forest resources and forestry and making Forest Operation Plans. The associations, however, have also been weakened by the decline in domestic forestry and the lack of forestry labour. The Forestry Agency has promoted the amalgamation of forest owners' associations based on the Forest Owners' Association Amalgamation Promotion Law of 1963. One major policy objective is to maintain forestry productivity and retain forestry labourers by amalgamating weak associations. Frequently, the jurisdiction of amalgamated associations includes territory in several municipalities. This has altered the direct relationship between associations and municipalities. It is important to note that such a program has been promoted at the same time that departments specific to forestry and staff with forestry education have virtually disappeared from municipal offices. This has led to the problem of who should be responsible for forestry administration at the municipal level.

One important forestry-related policy change in the 1990s was the increased role of municipal forest policy. The forest planning system began including a municipal component after the 1983 amendment of the Forest Law. With the 1998 Amendment, all municipalities must make a Municipal Forest Plan that includes all forest practices from planting to final cutting. In the 1998 Amendment, the responsibility for several forest practices, such as accepting harvesting reports, approving Forest Operation Plans, and making administrative recommendations for forest practices, was moved from prefectural governments to municipal offices.

In a questionnaire distributed to municipal offices in Kagoshima Prefecture, Matsushita (1999) asked, "What do you think of the transfer of responsibility for several forest practices relating to the forest planning system in the 1998 Amendment?" The responses from a total of 90 municipalities were: "It is very difficult with the current staff and budget" (25.6%), "It is difficult with the current staff and budget" (40.0%), "It is possible with the current staff and budget" (10.0%), and "I don't know" (24.4%). The problem of who will make Municipal Forest Plans and administer forest practices after these responsibilities have been transferred from the prefectural government to municipalities must be discussed. From this perspective, a new relationship between forest owners' associations and municipalities must be constructed to develop a municipal-level forestry program.

In any discussion of rebuilding the relationship between forest owners' associations and municipalities, the policy of amalgamating forest owners' associations must be considered carefully. This policy has been promoted since the 1963 legislation, and has been strongly promoted recently in connection with the Forestry Agency's forest management policy based on river basins, which is based on the 1991 Amendment to the Forest Law. The basic concept behind this new policy is that promoting forest resource improvement and increasing forestry production are the responsibility of everyone involved in forest resources and forestry within a river basin, which becomes the planning unit of a Regional Forest Plan and includes several municipalities. The Forestry Agency divided all forestland into forest planning areas based on river basins, and introduced a program of forestry development based on the new planning units. The policy target of amalgamation of forest owners' associations is therefore amalgamation at the level of a forest planning unit that includes several municipalities. It is clear that the forest owners' association is still the lowest level of forestry administration in the amalgamation policy. Because postwar forest owners' associations were developed as municipal units, however, and the difference between the administrative areas of these associations and municipal forestry administration is not clear, the current unified amalgamation policy may cause other problems.

Finally, since the forest owners' associations serve as the lowest level of forestry administration, the problem with their legislative position must be pointed out. Under the current Forest Law, such associations are both public interest organizations and cooperatives. The former characteristic does not conflict with an administrative role in general, but since cooperatives are basically connected to private interest, the position of the associations is not always equivalent to that of an administrative organization. For example, conflicts may arise in the specification of protected and recreational areas that are strictly related to the interests of people other than the forest owners. The ambiguous legal position of forest owners' associations also contributes to the complicated relationship between associations and municipalities.

Forest Owners' Associations and Forest Management

An increasing number of forest owners have abandoned forest management because of the difficult economic conditions surrounding forestry, which include the aging of forest owners and sluggishness in long-term timber prices. Given such conditions, the idea that forest owners' associations ought to conduct forest management on behalf of forest owners has been suggested recently. The government has just begun to consider amending the Basic Forestry Law, the most basic legislation related to forestry policy.

As a first step, a report entitled *Basic Policy Committee for Forests, Forestry, and the Forest Products Industries – Basic Problems with Forests, Forestry, and the Forest Products Industries* was released in July 1999 (Forestry Agency 1999c). The report discussed the necessity of clarifying the measures that various types of forest managers, including forest owners' associations and forest owners, can take to manage forests. The committee has high expectations for forest owners' associations to manage forests on behalf of owners who cannot. Forest management was included in the list of businesses that forest owners' associations can conduct (Article 26 of the Forest Owners' Association Law). Provisions for forest management by forest owners' associations were added in the 1974 Amendment to the Forest Law, to counter the increase in absentee forest owners. At present only a few forest owners' associations manage forests (Hirata 1996; Matsushita 1998).

Initially, attention must be focused on forest owners' opinions of forest management by the associations. In a survey conducted for representatives of the associations in the Osumi Forest Planning Area of Kagoshima Prefecture (Matsushita 1997, 1998), 50.3% of 793 forest owners answered that they had no desire to entrust forest management to the forest owners' association. In contrast, 8.9% answered that they would entrust forest management to the association unconditionally, and 30.4% answered that they might entrust forest management to the association if there were no cost involved.

It is important to note the reasons why almost half of the forest owners were unwilling to entrust forest management to an association. A total of 370 owners were allowed to make multiple responses. The main reasons were: "I want to manage it by myself while I can" (60.5%), "I want a free hand in management by myself" (53.0%), "I can manage it by myself because the area is so small" (34.6%), and "I want to manage it by myself for purposes of pleasure or health" (31.4%). Although most forest owners are old, they want to manage their forests themselves as much as they can, regardless of whether there is anyone to succeed them. In other words, they have no desire to entrust the forest practices necessary in their forest to others. The fact that 30.4% would agree to entrust forest management if no cost were involved also clearly illustrates the attitude of forest owners: most focus on the ownership of forestland rather than forestry production. This has been clarified in many recent questionnaires. Overall, very few forest owners want to entrust forest management to the forest owners' associations.

Another problem with management by forest owners' associations is whether the associations can manage the entrusted private forest effectively. The aging of forest owners is one of the main reasons that they cannot continue forest practices; the associations' forestry workers are also aging, so how can the associations obtain the necessary forestry labourers?

A more serious problem is related to the money needed to manage forests. In most areas, forest owners and their families do the necessary work in private forests, and this does not require the direct expenditure of money. When a forest owner entrusts forest management or individual forest practices to an association, however, he has to pay both the real cost and commissions to the association. As mentioned earlier, owners do not want to pay additional money for entrusting forest management to an association. An investment in forestry does not interest forest owners because of the long-term sluggishness in timber prices.

Even if the problems with obtaining forestry labour and management costs are solved, it remains to be determined how forest owners' associations will benefit from the entrusted forest. If associations can provide a forest management service, with the necessary labour force and with sufficient subsidies for management costs from national and prefectural governments, forest owners may entrust their forests to them, especially if the forest is inconveniently located. At that point, however, the forest owners' associations will quickly face managerial difficulties.

After the Basic Forestry Law was enacted in 1964, it was expected that forest owners' associations would conduct forest practices on behalf of forest owners, and the Forestry Agency continues to subsidize the associations based on this expectation. The associations are the lowest level of forestry administration, and at the same time they can get subsidies for the group of forest owners. Whether forest owners' associations can really manage their members' forests must be examined in more detail.

Forest Owners' Associations and Forest Owners

Since a forest owners' association is essentially an organization of the forest owners within its area of jurisdiction, the relationship between the association and the forest owners – its members – is of particular importance. There are several problems with this relationship, which should be discussed: (1) the promotion of amalgamation of forest owners' associations, (2) changes in the types of business engaged in by the associations, (3) Forest Operation Plans formulated by the associations, and (4) changes in the characteristics of forest owners as examples of problems in the relationship between associations and forest owners.

Amalgamation of Forest Owners' Associations
The Forestry Agency has long promoted the amalgamation of forest owners' associations in order to strengthen their management structure. With the introduction of the forest management system based on river basins in the 1991 Forest Law, the Forestry Agency has promoted the amalgamation of the forest owners' associations within a river basin, which are the unit areas for Regional Forest Plans. Such a wide amalgamation of forest owners' associations would seem to create an excessive distance between the associations and forest owners. There has been no survey at the national level of how forest owners regard this amalgamation.

In a survey conducted in the Osumi Forest Planning Area of Kagoshima Prefecture, Matsushita (1997) asked, "As a forest owner, what do you think of the wide amalgamation of forest owners' associations?" The responses were: "Amalgamation should definitely go ahead" (15.4%), "Amalgamation should progress gradually by common consent" (19.8%), "Amalgamation counters forest owners' interests" (24.1%), "I don't know" (32.2%), and no response (8.6%) (total responses = 793). Excluding the last two, the most common response was that "amalgamation counters forest owners' interests."

Legislative bodies cannot ignore the fact that many forest owners view the amalgamation of forest owners' associations negatively or critically. Forest owners are afraid that their relationship with the associations will become estranged with amalgamation. The associations will have to determine whether the managerial strengthening resulting from amalgamation is preferable to the close relationship that currently exists between them and forest owners.

Amalgamation is strongly directed by the Forestry Agency, however, and has been promoted throughout Japan. Furthermore, the Forestry Agency has determined that the forest owners' association is the lowest-level organization for forestry administration. Associations cannot determine whether to amalgamate by themselves, and amalgamation is mainly the result of the administrative policies of the prefectural forestry departments and the Forestry Agency. Consequently, the relationship between associations and forest owners appears threatened by amalgamation. The Forestry Agency (2000) has stated that forest owners' associations are expected to act as a coordinator between the administrative sector and forest owners, but it is doubtful whether the associations can play such a role after amalgamation.

Businesses Engaged in by Forest Owners' Associations
The business of forest owners' associations has been changing, and the change has brought about a change in the relationship between the associations and forest owners. Since the Basic Forestry Law was enacted, the Forestry Agency has strongly promoted the planting of needle-leaved trees throughout the country, and forest owners' associations have played an important role in the planting program. Reforestation, the most important business of the associations, has grown in tandem with the government reforestation program, the major objective of which is to increase planting of needle-leaved trees to produce construction wood.

Associations and forest owners used to have a very close relationship because many forest owners could obtain planting subsidies through their associations and were then able to complete planting and initial tending, such as weeding, themselves. Now many forest owners have negative opinions about forest management practices such as planting, thinning, and cutting. The forests planted during the 1960s and 1970s need thinning, but many owners are unwilling to thin their forests given the current price of timber. Furthermore, many owners of forests in which the trees have reached the age of final cutting according to their Forest Operation Plans have avoided cutting because the expense of subsequent replanting makes cutting unprofitable. Consequently, the area of new planting has been decreasing. Many forest owners are also unwilling to replace broad-leaved trees with needle-leaved trees, and so the associations have been unable to increase their volume of business with private forest owners.

Under these circumstances, the associations depend on business generated by management of public forests, including municipal forests, and on afforestation conducted by the Forest Development Corporation and prefectural afforestation corporations. Since the associations can rely on an almost constant annual volume of business from the public sector, which is managed under an annual budget system, the public sector is contributing increasingly to the managerial stabilization of forest owners' associations. In contrast, since most private forest holdings are small and the volume of forest practice varies, it is difficult for associations to obtain a constant volume of business from private forest owners. As associations tend to favour forest management practices related to the public sector rather than to private forest owners, from the perspective of the owners their associations do not play an effective role as representative organizations. Furthermore, considering that associations deal with forest management practices involving various non-timber utilities, as set forth in the Forest Owners' Association

Law, the relationship between them and private forest owners must be reconsidered.

Forest Operation Plans Formulated by Forest Owners' Associations
The 1968 Amendment to the Forest Law said that Forest Operation Plans were to be made by private forest owners. The 1974 Amendment introduced the Territorial Joint Forest Operation Plan, in which multiple forest owners can make a Forest Operation Plan under specific conditions, and now most Forest Operation Plans are joint plans made by forest owners' associations on behalf of forest owners.

When a forest owners' association makes an initial Territorial Joint Forest Operation Plan, the relationship between the association and the forest owners is very important. Forest Operation Plans must be revised every five years, according to the Forest Law. In most cases, this tends to be a formality, without an on-site survey. For example, Matsushita (1997) conducted a survey in the Osumi Forest Planning Area of Kagoshima Prefecture, where over 90% of the non-national forests are covered by Forest Operation Plans. Out of 710 responses, 48.9% of forest owners answered that they did not have a Forest Operation Plan, and 38.3% answered that they did not know whether they had one. This indicates that the forest owners' association generally made Territorial Joint Forest Operation Plans without the agreement or cooperation of the forest owners, further evidence of the increasingly estranged relationship. Given this situation, it is also doubtful whether the forest practices set out in the Forest Operation Plans of most forest owners' associations will actually be implemented.

Changes in Characteristics of Forest Owners
Typically, many forest owners are farm households that own forest. It is important to note that forest owners' associations are organizations of forest owners, and not organizations of people involved in real forest practices, forest production, or wood-based industries. The main reason most private forest owners continue to own forestland is to retain ownership of their ancestral forestland rather than to provide a continuous supply of timber. On the other hand, the major objective of the subsidies provided by the Forestry Agency is to realize appropriate forest management and promote forestry development, not to support the individual property management of forest owners.

Before the economic boom, forest resources contributed to the income of farm households to some degree, but this contribution has gone down drastically. The value of standing trees has decreased with the fall in domestic log prices, and has reached the point where many forest owners have given up positive forest practices, while retaining ownership of forestland as real estate. Many forest owners are unwilling to sell their forestland, even though they have stopped carrying out necessary forest practices such as thinning, because they consider the forest property part of their homestead. In this sense, one characteristic of current forest owners' associations is that they are organizations of people who simply own forestland. Furthermore, this trend is strengthening, and government policy on forest owners' associations must therefore be reconsidered.

The report entitled *Basic Policy Committee for Forests, Forestry, and the Forest Products Industries: Basic Problems with Forests, Forestry, and the Forest Products Industries* pointed out that the forest inheritance tax policy related to private forest owners must be reconsidered with careful attention to the equitability of taxation. In short, the report proposed reducing the forest inheritance tax, because the worsening economic situation of private forest owners makes it difficult for them to continue long-term forest management. There is, however, no clear reason why forest owners should be made a special case in the inheritance of private property simply because they own ancestral forestland. More attention must be focused on the obstacle that the continuous holding of forestland as household property represents to the development of private forestry in Japan, since most forest owners have little desire to produce timber continuously or to manage their forests effectively. Whenever the basic policy on forest owners' associations is reconsidered, the responsibilities of forest owners – the members of those associations – must be discussed simultaneously.

References

Forestry Agency (1997a). "Shinrin kumiai seido kentokai hokoku," *Rinya Jiho* (511): 28-31 (in Japanese).

– (1997b). "Shinrin kumiai ho oyobi shinrin kumiai gappei jyosei ho no ichibu kaisei no haikei to gaiyo," *Rinya Jiho* (514): 25-29 (in Japanese).

– (1998). *Ringyo hakusho,* fiscal 1997 edition, 66-68. Nihon Ringyo Kyokai, Tokyo (in Japanese).

– (1999a). *Shinrin kumiai tokei,* fiscal 1997 edition, 322-23. Zenkoku Shinrin Kumiai Rengokai, Tokyo (in Japanese).

– (1999b). *Ringyo hakusho,* fiscal 1998 edition, 77-79. Nihon Ringyo Kyokai, Tokyo (in Japanese).

– (1999c). "Shinrin ringyo mokuzai sangyo kihon seisaku kentokai hokoku," *Rinkeikyo Geppo* (455): 4-21 (in Japanese).

– (2000). *Ringyo hakusto,* fiscal 1999 edition, 69-70. Nihon Ringyo Kyokai, Tokyo (in Japanese).

Fujita, Yoshihisa (1998). "Kakuitsuteki na shinrin kumiai koikika heno gimon" ("Questioning uniformly widening forest owners' associations"), *Ringyo Keizai (Forest Economy)* (600): 22-24 (in Japanese).

Handa, Ryoichi, and Sumiyoshi Ariki (1996). "Shinrin kumiai seido no hensen." In Handa, Ryoichi, ed., *Rinseigaku,* 2nd printing, 70-72. Bun-eido Shuppan, Tokyo (in Japanese).

Hirata, Kunihiro (1996). "Shinrin kumiai ni yoru chiiki no shinrin kanri ni kansuru kenkyu" ("Study on forest management in an area by the forest owners' association"), *Ringyo Keizai Kenkyu (Journal of Forest Economics)* (129): 159-64 (in Japanese).

Kaibara, Ichiro (1955). *Ringyo seisaku ron*, 432-35. Rinya Kosaikai, Tokyo (in Japanese).

Kaneiwa, Kunio (1953). "Sanson keizai ni okeru shinrin kumiai no tachiba." In Nogyo-To-Keizai Hensyubu, ed., *Rinya to nogyo keiei*, 46-57. Kawasaki Shuppansha, Tokyo (in Japanese).

Matsushita, Koji (1995). *Ringyo koyo kaizen sokushin jigyo kenkyukai hokokusyo (Kagoshima-ken Hokusatsu ryuiki)*, 24-40. Kagoshima-Ken Shinrin Kumiai Rengokai, Kagoshima (in Japanese).

– (1996a). "Recent problems of the forest insurance system in Japan," *Finnish Journal of Business Economics* 45(1): 86-100.

– (1996b). "Ringyo rodosha no koyo joken ni kansuru kenkyu – shinrin kumiai sagyohan no chingin suijyun –" ("An analysis of forestry workers' employment conditions – wages of the forest owners' associations in the Hokusatsu Forest Planning Area of Kagoshima Prefecture –"), *Kyoto Daigaku Nogakubu Enshurin Hokoku (Bulletin of the Kyoto University Forests)* (68): 62-76 (in Japanese with an English summary).

– (1997). *Ringyo koyo kaizen sokushin jigyo kenkyukai hokokusyo (Kagoshima-ken Osumi ryuiki)*, 8-18. Kagoshima-ken Shinrin Kumiai Rengokai, Kagoshima (in Japanese).

– (1998). "Shinrin kumiai heno keiei itaku ni kansuru shinrin shoyusha no ikou" ("Forest management by local forest owners' associations"), *Kyoto Daigaku Nogakubu Enshurin Hokoku (Bulletin of the Kyoto University Forests)* (69): 54-67 (in Japanese with an English summary).

– (1999). *Ringyo koyo kaizen sokushin jigyo kenkyukai hokokusyo (Saishu-nendo ban)*, 40-54. Kagoshima-ken Ringyo Ninaite Ikusei Kikin and Kagoshima-ken Ringyo Rodoryoku Kakuho Shien Center, Kagoshima (in Japanese).

Shinrin Kumiai Seido Kenkyukai (1998). *Meikai Shinrin kumiai seido no kaisetsu*. 302 pp. Zenkoku Shinrin Kumiai Rengokai, Tokyo (in Japanese).

Shinrin Kumiai-ho Kenkyukai (1989). *Shinrin kumiai no jitsumu – ichimon itto shu –*, 21-23. Nippon Ringyo Chosakai, Tokyo (in Japanese).

Shioya, Tsutomu (1973). *Rinseigaku*, 165-80. Chikyusha, Tokyo (in Japanese).

Tanaka, Shigeru (1996). "Shinrin-kumiai." In Handa, Ryoichi, ed., *Rinseigaku*, 2nd printing, 181-90. Bun-eido Shuppan, Tokyo (in Japanese).

Zenkoku Shinrin Kumiai Rengokai (1973). *Shinrin kumiai seido shi*, 3: 486-88. Zenkoku Shinrin Kumiai Rengokai, Tokyo (in Japanese).

5
Forestry Labour
Ichiro Fujikake

Japan employs one of the most labour-intensive silvicultural practices in the world. Several reasons have been advanced for this:

- Because the mild and humid climate in most parts of the country allows bushes and vines to grow vigorously, a fair amount of work, including careful site preparation, planting, and weeding, is required to permit productive species to grow under favourable conditions.
- As part of the traditional national culture, the mode of housing construction provides the basis for people's unique preference for construction lumber that is often produced from timber using intensive silvicultural management. Dense planting coupled with subsequent intensive thinning and pruning, both often aimed at producing decorative pillars used in housing construction, are good examples of labour-intensive techniques characterizing Japanese forestry.
- The dense population sustained by rice production, which is said to be capable of supporting a significantly higher population density, has provided a rich source of forestry labour supply in rural areas, especially before the high economic growth following the Second World War.

Today, one more reason for the intensive labour practices of Japanese forestry should be added, namely, lower mechanization of silvicultural projects compared with other countries because of Japan's mountainous topography.

Reflecting the intensive labour involved in Japanese forestry, research on forestry labour in Japan has a significant share in the forest economics literature. In particular, many authors have focused on poor employment conditions and the resulting low welfare at forestry workplaces, and the steady decline in the number of forestry workers; both are the consequences of the declining competitiveness of domestic forestry relative to foreign exporting countries, and of changes in the general structure of the Japanese economy.

The aim of this chapter is to introduce some aspects of Japanese forestry labour after the Second World War based on the forestry labour economics literature up to the present day. The first section, "Demand and Supply of Forestry Labour, and Related Policies after the Second World War," attempts to explain the performance of the forestry labour market and related policy measures in the postwar period. The second section, "The Current Japanese Forestry Labour Market," discusses topics that are thought to be important in understanding today's forestry labour market and its prospects, focusing particularly on the supply side of forestry labour. Specifically, forecasting of labour supply, wage rates, attitudinal changes towards forests and forestry by the public, and the quality of forestry labour will be discussed.

A few remarks are now in order about the scope of the discussion here. Both employees and self-employed workers are included in the definition of forestry workers. Regarding categories of work, the discussion does not deal with workers in sawmills, pulp mills, and so forth, but rather with workers in silvicultural activities and logging. In practice, silvicultural workers and loggers are usually differentiated, but they are considered together here unless otherwise stated.

Demand and Supply of Forestry Labour, and Related Policies after the Second World War

The Japanese economy grew very rapidly after the Second World War, at an average yearly rate of 11% from the latter half of the 1950s into the 1970s. Japan also experienced rapid economic growth at an annual rate of 9% from 1946 to 1955, but this is usually distinguished from growth in the period that followed in that it depended on demand created by the postwar economic recovery and demand from the Korean War on the consumption side, and the postwar resumption of production on the supply side. The high economic growth began with plant investment for modernizing production facilities, and it accelerated the modernization of Japan in terms of industrial structure and consumption behaviour. The growth and transformation of the general economy during this period had, of course, an enormous influence on the forestry labour market.

Before looking at the changes in demand and supply in the labour market during the period of high economic growth, we will discuss the forestry labour market during the period of recovery following the war.

Postwar Recovery Period, 1946-54

During this period, Japan had a large rural population in agricultural and forestry villages despite the fact that the population drain from rural to urban areas had already begun, as indicated by a drop in the rural population from 63% in the 1950 census to 44% in the 1955 census. In particular, before the demand resulting from the Korean War stimulated the Japanese economy, the farm economy in rural areas appeared to have played an important role in sustaining overpopulation due to the repatriation of soldiers and citizens after the Second World War (Miyazaki and Ito 1989). Because not enough non-agricultural and non-forestry jobs were available in rural areas at that time, the rural population was a large potential supply of forestry labour. At the same time, the demand for timber was quite high because of both the postwar recovery and the Korean War. The high demand stimulated not only harvesting but also planting. The increase in timber prices in this period provided a strong incentive for planting.

Figure 5.1 shows the ratio of timber prices (yen/100 m³) for two main commercial species, sugi (Japanese cedar, *Cryptomeria japonica*) and hinoki (Japanese cypress, *Chamaecyparis obtusa*) to the daily wage of a logging crew (yen/worker-day). The value of timber relative to forestry wages peaked in 1960, encouraging forest owners to plant; in such economic conditions, forest owners could employ forestry labourers to plant and maintain their forests without severe economic burden. Furthermore, since the so-called energy revolution (the switch to coal, oil, and gas as energy sources) had not yet occurred, fuelwood production employed a significant labour force at that time.

Thus, because of favourable conditions in both the demand and supply sides, the number of workers employed in the Japanese forestry sector reached

Figure 5.1

Number of loggers (worker-days) affordable per 100 m³ based on stumpage prices, 1950-90.

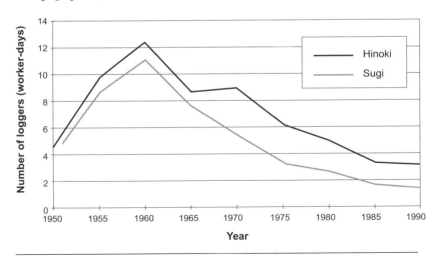

Sources: Japan Real Estate Institute, *Forest Land Price Index;* Ministry of Labour, *Wage Survey of Forestry Workers by Occupation,* 1950-90.

Table 5.1

Number of forestry workers in Japan, 1955-95.

Year	Number of forestry workers (thousands)	Rate of decline (%)
1955	519	–
1960	439	15
1965	262	40
1970	213	19
1975	179	16
1980	165	8
1985	140	15
1990	108	23
1995	86	20

Source: Prime Minister's Office, Population Census, 1955-95.

a historically high level during this period. There are three sources of statistics relating to the number of workers employed in forestry: the population census, the employment status survey, and the labour force survey. Only the population census has statistics for both before and after the Second World War. According to the data, the number of workers in forestry reached a peak of 519,000 in 1955, as shown in Table 5.1.

Careful analysis of the employment status of forestry workers reveals that only a small proportion were full-time and organized into formal organizations. Many workers were part-time, self-employed, or employed within the family, typically without a formal employment contract. From the standpoint of forest owners, forestry management during this period depended greatly on family labour input. According to the world census of agriculture and forestry in 1960 (Ministry of Agriculture and Forestry), forest owners (defined as having more than 1 ha of forestland) input 57 million worker-days, with 78% consisting of family labour. Unfortunately, comparable statistics are not available for recent years, but the 1970 census showed that, while total labour input had dropped to 27 million worker-days, the percentage of family labour had already dropped to 65%. It is clear that before Japan experienced high economic growth, family labour was by far the most important source of forestry labour supply.

To understand the significance of family labour and labour relatively free from organization, such as part-time or self-employed labour, one must understand that forestry jobs absorbed surplus labour from farm households at that time. Farms were labour-intensive in planting and harvesting because of low mechanization, and in off seasons farm household members were unproductive. Forestry jobs are suitable for farm households as a source of surplus seasonal employment (University of Tokyo Social Science Institute

1954, 283-90). Allocating time to both farm and non-farm jobs was typical in rural areas, and a small proportion of non-farm jobs was in forestry. Managing family forests or being employed on a day-to-day basis when there were no farming activities resulted in family labour and labour relatively free from organization.

During this period, there were no labour laws specific to forestry, but rather laws relating to general labour policy aimed at democratization, including the Labor Standard Law and the Union Law (Nonomura 1983). Because there were a sufficient number of workers in forestry, special forestry labour policy was not necessary. This, however, was not the case in succeeding periods.

Period of High Economic Growth (1955-72)

In the period of high economic growth, the development of the general economy brought about an increasing productivity gap between primary industry and other industries, and consequently a change in industry structure. This had a tremendous effect on the forestry labour market. There was a rapid drop in both workers in primary industry and the rural population. In addition, two forestry-related factors at the beginning of this period led to a dramatic decline in the number of forestry workers.

First, as a result of the energy revolution, firewood production, which had occupied a large share in forestry, declined by 50% from 1957 to 1962. Second, although the demand for sawnwood timber continued to rise during the period of high economic growth, around 1960 the government implemented a set of policy measures to promote timber imports in order to ease the tight timber market and stabilize timber prices. This shifted the demand for timber from domestic to lower-cost imported timber. As a result, prices for domestic timber continued to decline relative to forestry wages, as shown in Figure 5.1, and supply and demand for forestry labour decreased rapidly in the 1960s. Table 5.1 shows the decline in the number of forestry workers in this period.

It should be noted that this decline was preceded not only by an outflow of the rural population to urban areas but also by a change in industry structure in rural areas. Since relocation of employees and employers is costly, higher urban wages do not necessarily attract all rural workers. The deep-rooted practice of inheritance where the eldest son inherits everything forced at least the eldest son to remain in the village. These observations show the regional independence of the labour market. In light of this local phenomenon, labour markets in urban and rural areas are independent of each other to some extent, and industries or employers compete in the regional labour market. At the beginning of the period of high economic growth, secondary and tertiary industries increased production, and consequently employment, almost exclusively in urban areas. As the economy developed further,

Table 5.2

Number of forestry workers in Japan by employment status, 1956-87.

Year	Self-employed (thousands)		Family workers (thousands)	Employees (thousands)		
	With employees	Without employees		Permanent	Temporary	Daily
1956	9	104	63	113	38	85
1968	9	33	20	100	59	30
1977	9	20	11	89	16	41
1987	7	17	10	65	9	22

Source: Prime Minister's Office, Employment Status Survey, 1956, 1968, 1977, 1987.

however, industries, mainly manufacturing, began to expand their operations to rural areas, seeking cheap labour. This resulted in the broadening of occupational choices in rural areas and a consequent decline in the competitiveness of forestry in the rural labour market (Hamada 1982).

Table 5.2 summarizes the employment status of forestry workers. The first two rows show a sharp decrease in self-employment or family employment, a slight decrease in full-time employment, and an increase in part-time employment. This indicates a shift in the employer/employee relationship in forestry towards a formal and permanent one, an extension in workdays, and a shift to full-time employment. These changes are closely related to changes in farm management (Okamori 1979). As the general economy developed, both domestic agriculture and forestry lost competitiveness. Many farm households abandoned agriculture, but most chose a way in which they reduced their reliance on agriculture for household income and came to rely instead on deriving income primarily from non-agricultural jobs. As industries established factories in rural areas and motorization developed, more and more young and middle-aged workers chose non-agricultural occupations as their primary occupations. Mechanization of agriculture and the popularity of agricultural chemicals during this period helped this trend by reducing the time needed to cultivate land (Inoki 1989) and making it possible to rely on agricultural work performed mostly by women and the elderly. The increased numbers of farm household members employed full-time in non-agricultural industry resulted in a drop in the forestry labour supply.

Changes in occupational choice in rural labour markets were observed during the period of high economic growth. Figure 5.2 presents an explanation of typical occupational choices in both the postwar recovery period and the period of high economic growth. The horizontal axis represents time input from right to left. Excluding leisure time for the sake of simplicity, it is assumed that a worker maximizes income by allocating the time t_0-t_4 to various jobs. The vertical axis represents income. The slopes of F1

Figure 5.2

Theoretical explanation of the occupational choices of farm households in Japan during the postwar recovery period and the period of high economic growth.

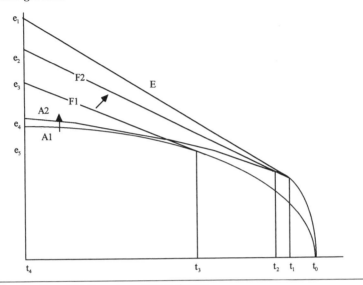

Note: An explanation of this figure is provided in the text.

and F2 represent the wage rate of forestry jobs; E represents the wage rate in manufacturing factories; and A1 and A2 represent the rate of earnings from agriculture during the recovery period and the period of high economic growth, respectively.

Because agriculture uses typically self-employed rather than paid labour, income per unit time cannot be represented by a constant wage rate. Assuming that household land, and thus land input by a household, is fixed, the marginal value of labour decreases with labour input. To put it simply, forestry work on family forests is excluded. It is also assumed that there were no manufacturing factories during the recovery period. Reflecting the small improvement in labour conditions in forestry, the slope of F2 is only slightly steeper than that of F1. On the other hand, because of mechanization and chemical use in agriculture during the growth period, high labour productivity was achieved with smaller labour input. Because of fixed land input between periods, however, labour productivity decreased rapidly as labour input increased during the later period.

In these conditions, occupational choice is optimized during the recovery period by allocating t_0–t_3 to agriculture and t_3–t_4 to forestry. This occupational choice generates greater income than when the whole time is devoted to agriculture ($e_3 > e_4$). In the period of high economic growth, factories were introduced to rural areas and provided higher wages than forestry (the slope of E is greater than that of F2), whereas in the agricultural sector, farmers were able to fully utilize their land with a small labour input. As a result, the time allocated to agriculture is reduced to t_0-t_1 and the time then allocated to forestry has to be reallocated to paid work in a factory. This occupational choice increases total income to e_1, which is larger than e_2, the income achieved by a combination of agriculture and forestry, which was typical in rural areas during the period of high economic growth. Yet there were regions where non-agricultural work was not available, and sometimes non-agricultural employers did not employ the elderly. In these situations allocating t_0-t_2 to agriculture and t_2-t_4 to forestry is the best solution. Workers in such situations were in fact the main source of forestry labour supply during that period.

Turning to forestry labour policy, as the large-scale outflow of labour to urban areas began during the period of high economic growth, forest owners and other stakeholders began to seek public assistance in improving forestry labour conditions to counter the rapid decline in forestry labour. Concrete policy measures were implemented after 1965 in accordance with the philosophy of the Forestry Basic Law, which aimed at the development of forestry as an industry. Corresponding to changes in the economic environment surrounding forestry labour markets, forestry labour policy during this period emphasized changing the occupational status of forest labourers from part-time to full-time in order to prevent rural labourers from changing to full-time paid employment in industry. Forestry cooperatives played an important role in forestry labour policies. Such policies aimed at encouraging workers to enter forestry jobs were linked to the idea of regional co-operation in forestry. Specific policy measures included the increase and concentration of work under forest owners' associations using subsidies for various forestry practices, and the improvement of labour conditions in forest owners' associations were considered necessary in achieving full-time employment (Izumi 1991).

Although there is no doubt that the policies to attract forestry workers helped improve labour conditions, they fell short of achieving parity between forestry labour conditions and those in secondary and tertiary industries in general. As a consequence, although the forestry sector succeeded in securing elderly labour, and to some extent in promoting full-time employment, younger workers flowed out of forestry seeking better labour conditions. Thus the change of generation in forestry labour was not completely realized. This had a serious effect on the forestry labour market in the lower-growth period that followed.

Period of Low Economic Growth (1973-present)

The rapid growth of the Japanese economy came to a halt during the first

Table 5.3

Number of forestry workers in Japan by age, 1955-95.

	Number of workers (thousands)		
Year	15-34 years	35-54 years	≥55 years
1955	267	181	71
1965	95	123	43
1975	26	112	41
1985	14	74	52
1995	8	29	49

Source: Prime Minister's Office, Population Census, 1955-95.

Table 5.4

Net increase of forestry workers in Japan by cohort, 1955-95.

	Number of workers (thousands)[1]				
Period	15-24 years	25-34 years	35-44 years	45-54 years	55-64 years
1955-65	-59	-63	-49	-53	-39
1965-75	-7	-14	-18	-18	-20
1975-85	4	1	-2	-17	-18
1985-95	2	0	-2	-20	-27

1 Age ranges indicate ages of the cohorts at the beginning of the period.
Source: Prime Minister's Office, Population Census, 1955-95.

oil crisis at the beginning of the 1970s, and the economy entered a period of low growth. The decline in forestry labour continued. As shown in Table 5.1, the decline slowed between 1975 and 1980 but accelerated again because of aging of the forestry labour pool that had taken place over a number of years.

Table 5.3 shows that there was significant aging during the period of low economic growth. While the proportion of forestry workers aged 55 and over rose by only 3% from 1955 to 1965 and 6% from 1965 to 1975, it rose by 14% from 1975 to 1985, reaching 37%. This was because a large cohort that had entered the forestry sector in their youth during the recovery period had now reached an advanced age.

Reflecting this, the way in which forestry labour declined differed between the high- and low-growth periods. Table 5.4 shows changes in the size of forestry labour cohorts. It is clear that the main factor in the decline in forestry labour was the change from young generation to older generation year after year. It can be safely assumed that leaving a forestry job at age 65 or over means retirement from the labour market. Depopulation in Japanese rural areas is said to have reached a stage of natural decline as opposed to social decline, but the same also holds true for forestry.

According to the most recent population survey, in 1995, there were about 86,000 forestry workers, representing a decline to one-fifth of the peak number in 1955. Of these, 57% are over the age of 55, so a further decline in the near future is unavoidable. Female workers account for 17% of the total. Typical female forestry workers are farm household members who are hired, mostly by forest owners' associations, as forestry employees only in agricultural off seasons to supplement male workers. The type of work that female workers do is characteristically low risk, including planting and manual weeding, and in general the wage rate is quite low relative to male workers (Makino 1986). Full-time employment became more common in the period of low economic growth, as shown in Table 5.2. No data are available categorizing forestry workers as a whole by type of work engaged in, but the Statistics of Forest Owners' Associations (Forestry Agency 1993), which deals only with workers organized into forest owners' associations, provides satisfactory data because such workers make up a large percentage of total forestry workers. According to the statistics, workers engaged mainly in planting made up 69%, logging 18%, and others 13%. Recent trends indicate a decrease in the rate of those engaged mainly in planting and an increase in the rate of those engaged in logging, because expansive afforestation has almost ceased throughout the country.

During the period of high economic growth, policy aimed at attracting forestry labour stressed full-time employment; it succeeded to some extent in keeping middle-aged workers in forestry but failed to attract younger workers. In the stage of natural decline in forestry labour, it is obvious that policies effective only in retaining the middle-aged are meaningless. Thus the focus during the period of low economic growth was on attracting the younger generation to forestry jobs.

In spite of gradual improvement, forestry labour conditions continued to be very poor compared with industry during the period of high economic growth. For example, the daily average wage for a logging crew was $1/40$ the monthly average income in industry throughout the postwar period (Basic Survey on Wage Structure and Wage Survey of Forestry Workers by Occupation, Ministry of Health, Labour, and Welfare, 1961-95). Wages for planting and other activities are generally lower than for logging. It is now clear that forestry wages rose at the same rate as industry, but remained lower relative to industry.

Policy measures introduced during the period of low economic growth led to dramatic improvements in labour conditions, especially wage increases to attract young labour in competition with other industries. Such measures were applied throughout the country in the 1980s, and included the introduction of monthly payments, sometimes with a fixed annual pay raise, housing services, and establishment of forestry employment agencies using both public and private funds. It was obviously not a simple matter, however, to offer wages at levels satisfactory to the younger generation when timber prices dropped to extremely low levels relative to wage rates. In order to pay the large wage burden and avoid an otherwise inevitable deficit, some kind of public support was introduced in most cases.

Because of the importance of public support, two notable features of forestry labour policy are evident in this period. The first is that employers eligible for such support were limited largely to forest owners' associations and forestry employment agencies. As a consequence, concentration of forestry labour under employers that are basically public in nature has occurred and will continue to occur. The second is that regions capable of adopting such policy measures requiring strong public support are limited to those with secure, positive support from municipalities. Forestry labour policy measures during the period of high economic growth were applied uniformly throughout the country. In the period of lower growth, however, some regions began to adopt more aggressive measures than others, and there appeared to be a trend towards widening regional differences in measures taken to attract young labour.

Other novel (for forestry) strategies for securing labour include advertising for workers in magazines or providing housing services, both of which are helpful in attracting young workers from urban areas. Such strategies have been more or less successful since their implementation at the beginning of the 1990s. As a result of population concentration in urban areas during the high-growth period, the population in rural areas declined to less than one-fourth the total population. The percentage of young people in the urban population was particularly high, so urban areas became a very attractive source of labour for a forestry sector in urgent need of young workers. In addition, the urban population became very sensitive to environmental problems, which increased interest in nature and the countryside, thereby enlarging the potential source of forestry labour. (Until the 1980s, applicants for forestry jobs were commonly from the same area as the employer.) Labour conditions in forestry have not changed markedly, so, taking into account cases in which workers take up forestry jobs after quitting jobs offering more attractive salaries and/or social security conditions, it is fair to say that the increase in forestry job applicants from urban areas was due largely to changes in values in the urban population.

The Current Japanese Forestry Labour Market

Recent Trends in Number of Forestry Workers, and Making Predictions

As discussed at the beginning of this chapter, silvicultural techniques in Japan are highly labour-intensive. Thus trends in the number of forestry workers and, especially after the Second World War, questions about how far that number would decline have been among the most important concerns for forestry economists and government officials.

As forestry workers are aging rapidly, and the rate of decline in their numbers is accelerating, estimating future numbers of forestry workers is crucial to making forecasts about Japanese forestry. Many attempts have been made to forecast the number of forestry workers over the next few decades primarily using cohort analysis. Given that there are certainly differences in labour market behaviour across generations of workers, the way that cohort analysis forecasts the number of forestry workers by applying a different moving rate for each cohort appears to be fairly reasonable. However, if forecasts made by Tanaka (1981), the Forestry Agency (1983), and Nagata and Terashita (1991) – all employing cohort analysis – are compared with actual values, it can be seen that all the forecasts have understated the rate of decline of forestry workers. This demonstrates the limitations of simply applying cohort analysis to making such forecasts. In addition, the Forestry Agency (1983) introduced a somewhat more complex method of forecasting. They first used cohort analysis to estimate the so-called potential labour supply; then used a resource table to estimate potential labour demand, and finally took a weighted average of potential demand and supply. In calculating potential labour demand for the future, they used several scenarios, but every scenario produced a potential demand that was far larger than the potential supply from cohort analysis, so that the resulting weighted averages all gave a far too optimistic forecast in terms of the decline in forestry workers.

One of the reasons that the rate of decline is underestimated when cohort analysis is used is that the net rate of leaving each cohort has been increasing over a number of years. This indicates that the situation is very serious in spite of all measures taken by employers and the government to attract forestry workers. Further, as can be inferred from inspection of Forestry Agency (1983) forecasts, the forestry labour shortage is, in fact, now quite severe. It can be reasonably assumed that the silvicultural activities assumed to be necessary in Forestry Agency (1983) scenarios have not been completed.

It appears that in order to be more accurate, forecasts of the number of forest workers need to be based on an understanding of the relationship between various social and economic factors and the trend in number of forest workers. For example, with regard to forecasts of the rapid aging of the Japanese population itself and the resulting labour shortage, Tanaka (1992) anticipates that competition between forestry and other industries in the labour market will become more intense. Forecasting forestry labour should not be done in isolation from the dynamics of the general economy. Unfortunately, however, attempts by the Forestry Agency (1983) to regress the number of forestry workers on socio-economic factors were unsuccessful, and there appear to have been no efforts made along such lines since.

There appears to be a new trend in the number of forestry workers: the number of individuals taking up forestry jobs upon graduation from secondary school has increased recently, in spite of a decline in total number of students, and thus the number of graduates taking jobs recently. According to the School Basic Survey (Ministry of Education, Culture, Sports, Science, and Technology 1992, 1993, 1996), the number of graduates who take jobs in forestry gradually recovered from a low of 124 in both the 1991 and 1992 school years to 187 in the 1995 school year. This phenomenon is considered a consequence of various measures to attract forestry workers. Tanaka (1994) released the results of a survey in which newcomers to forestry were asked for their evaluation of forestry as a place to work and so on. They showed that many newcomers consider forestry wages as average for the region. It may therefore be concluded that the strategies for attracting young workers have been successful to a certain degree. It is also possible, however, that the difficulty of getting jobs in industry because of the recession has increased the forestry labour supply.

Wage Rate and Regional Differences
Providing good labour conditions is the key to countering the forestry labour shortage. In particular, the wage rate is essential in determining the forestry labour supply. In a survey by Tanaka (1994), the most frequent answer to the question "What kind of improvement in particular is needed in order to encourage workers to take forestry jobs?" was "to introduce a monthly salary that provides a steady income."

Although improvements in labour conditions have been implemented in various regions, salary increases or the introduction of a monthly salary with fixed annual increments are often insufficient because of the future anticipated wage burden. As a result, forestry wages are low on the whole, and recently the rate of annual increase has been lower than in other industries. As mentioned earlier, the daily forestry wage is equal to about $1/40$ of the monthly salary in industry. After recovering to $1/35$ in 1975, it declined until 1990, when it reached $1/43$. This observation reflects pessimism about the forestry labour shortage overall, in spite of successes in securing labour in some regions.

As already noted, the gap between regions in terms of their capacity to offer higher wages will probably widen because of differences in general economic conditions and types of regional forestry practices. These regional wage differences will continue to be important factors affecting regional performance in securing labour. Two reasons for these differences are simply regional differences in general wage rates and the cost of living. Another reason may be differences between forestry regions in ability to pay higher wages.

These reasons appear to be supported by data. Table 5.5 shows the results of regression analysis in which harvesters' wages were regressed on

Table 5.5

Results of regression analysis of Japanese forest harvesters' wages in 1989 against annual farm household expenditure and timber price.

Variable	OLS estimate[1]	
Farm household expenditure (thousands ¥)	1.0430	(0.310)
Timber price (¥/m³)	0.0989	(0.023)
Constant	2048.3000	(1435.100)

Notes: Dependent variable: daily wage for loggers (¥)
$n = 27$
F (2,24) = 22.70
$R^2 = 0.75$
Adj. $R^2 = 0.63$
1 Standard error in parentheses.

two variables: annual farm household expenditure (in thousands of yen) (Ministry of Agriculture, Forestry, and Fisheries 1990a) and the ratio of gross value of timber harvested (Ministry of Agriculture, Forestry, and Fisheries 1990b) to volume of timber harvested (Ministry of Agriculture, Forestry, and Fisheries 1990b). The latter ratio is thought to represent the regional price level of timber.

The sample consists of observations from 27 prefectures for which harvesters' wages are published, and in which national forest is less than 40% of total forest area. The coefficients estimated for both variables are significantly positive at 1%. Two variables accounted for 2/3 of regional variation in harvesters' wages. The coefficient estimated for farm household expenditure is close to 1, indicating that if annual expenditure differs by ¥1 million between regions, regional daily harvesters' wages would differ by ¥1,000 on average. Assuming there are 200 annual labour days in forestry, a daily difference of ¥1,000 constitutes only ¥200,000 a year, which is far less than the difference in annual expenditure, i.e., ¥1 million. This suggests that forestry faces tough competition in the labour market in regions where household expenditure is high.

The estimate of the coefficient of the timber price is about 0.1, indicating that a ¥10,000 difference in timber price gives rise to a ¥1,000 difference in harvesters' daily wages. This is considered to be the consequence of differences in capacity to pay wages due to differences in the value of forest resource stock between regions. It is likely that such differences in capacity to pay wages will play a significant role in determining which regions can attract and keep forestry labour. This is not to say, however, that high-priced timber must be produced. Currently the fall in price of high-priced timber is particularly marked since less expensive glulam is being widely used. Areas of high-priced timber production may lose their capacity to pay higher wages in the future.

Changes in Attitude towards Forests and Forestry Work, and Volunteering

Wage rates are not the sole determinant of labour supply. Factors determining worker satisfaction include not only wages and other labour conditions but also factors beyond the control of employers, such as individual preferences for nature, rural living, or working in nature, which often depend solely on the person. Moreover, a person would be more willing to work in forestry at a lower wage if he or she had a more favourable reaction to forestry work, rural living, etc.

Recently the image of forestry among urban dwellers has begun to change. Given the increase in forestry work applicants from urban areas, the change is particularly interesting and appears to have had an influence on forestry labour supply. Although employers cannot control workers' individual preferences, it might be possible to reconsider details of work and working conditions and adapt them to suit those preferences. Mitsui (1994) argued that in order to encourage urban labour to take up forestry jobs, it is important to know why and how urban workers come to obtain forestry jobs, and to implement policies by modifying forestry techniques and rural living conditions accordingly. In fact, an enormous effort on the part of forest owners, forest owners' associations, and forestry officers at the regional level has been, and continues to be, devoted to these ends.

Aside from the entry of urban labour into forestry jobs, the increasing popularity of volunteer forestry work appears to be another consequence of the urban populations' changing attitudes towards forestry. In the latest nationwide survey conducted by the Prime Minister's Office, half of the respondents said that they would like to participate as volunteers in silvicultural work (Prime Minister's Office 1996). Although the amount of work performed by volunteers does not appear significant, volunteers of this type have never been seen in Japanese forestry. The popularity of volunteer forestry should be monitored.

Skill and Incentives to Work

The discussion so far has emphasized the quantitative aspects of forestry labour supply. Another important aspect is related to the quality of work. It is generally agreed that Japanese silvicultural techniques not only need high labour input but also require skilled labour. There are of course individual differences in skill. Nevertheless, some authors have conceived the idea that a worker's status or other socio-economic characteristics might affect the quality of work. It has been stated repeatedly that family labour or labour in a long-term relationship with a particular forest owner will be more efficient in some respects than casual labour, because the former have greater incentives to work than the latter, and the former are more efficient than the latter in the sense that a permanent relationship with a forest owner can provide the worker with local knowledge of forestland, management history, and other management-specific information (see Fujikake [1995] for a survey of this literature).

In terms of work quality, a decline in the skill of labour and a disincentive to work will be significant problems in the near future. Two points should be stressed.

- First, the organization of forestry labour under forest owners' associations is taking place with political support. Recently when there has been a labour shortage due to lower timber prices, it might have been efficient for forest owners to pool workers in forest owners' associations and use them collectively. It should be noted, however, that workers employed in this way might be inefficient for the reasons briefly alluded to earlier. It will be a challenge for forest owners' associations to overcome this weakness and make forest owners appreciate work performed by workers belonging to the associations.
- Second, while many forest owners do not actively manage forests, it is becoming difficult to transfer silvicultural techniques and management knowledge across generations in families. This situation is serious especially in the case of absentee owners, because once land is inherited, the heirs may not even know the boundaries of the land. Even owners living near their forests usually have occupations other than forestry, so the time available to a family for transferring techniques and knowledge is severely limited. Thus, maintaining the quality of family labour is an important problem today.

Training young workers is another problem Japanese forestry faces in sustaining the skills of forestry workers. In earlier days, forestry workers began at a young age and were gradually trained on the job. Today young workers enter forestry mostly after high school, without any forestry experience, although their wages are quite high from the outset. Efficient and effective training regimes for such workers are an important concern for forestry employers.

References

Forestry Agency (1983). *Kokusanzai kyokyu system keiryo model kaihatsu chosa houkokusho (Ringyorodo jukyu yosoku model)*. 139 pp. Forestry Agency, Tokyo (in Japanese).
– (1993). Statistics of Forest Owners' Associations, 1991 (in Japanese).
Fujikake, Ichiro (1995). "Ikurin seisan no ninaite no yakuwari ni tsuite." In Kitagawa, I., ed., *Shinrin, ringyo to chusankan chiiki mondai*, 159-74. Nippon Ringyo Chosakai, Tokyo (in Japanese).
Hamada, Mamoru (1982). "Rodoryoku shijo no juso kozo to ringyo rodo mondai" ("Several structures of labour market and the problems of forestry labour"), *Ringyo Keizai Kenkyu (Journal of Forest Economics)* (102): 23-27 (in Japanese).

Inoki, Takenori (1989). "Seicho no kiseki." In Yasuba, Y., and Inoki, T., eds., *Kodo seicho*, 97-151. Iwanami Shoten, Tokyo (in Japanese).

Izumi, Eiji (1991). "Ringyo rodoryoku no soshikika ni tsuite" ("On organizing forestry labour"), *Ringyo Keizai Kenkyu (Journal of Forest Economics)* (119): 28-37 (in Japanese).

Makino, Tomomi (1986). "Joshi ringyo rodosha no shugyo jotai to sekishutsu kozo ni kansuru kenkyu" ("Studies on the employment and making of female forestry labour. Two types of female work party in the forestry cooperatives"), *Ringyo Keizai Kenkyu (Journal of Forest Economics)* (109): 37-41 (in Japanese).

Ministry of Agriculture, Forestry, and Fisheries (1990a). Survey of Farm Household Economy, 1989 (in Japanese).

– (1990b). Survey on Lumber Production and Marketing, 1989 (in Japanese).

Ministry of Education, Culture, Sports, Science, and Technology (1992, 1993, 1996). School Basic Survey, 1991, 1992, 1995 (in Japanese).

Ministry of Health, Labour, and Welfare (1961-95). Yearbook of Labour Statistics, 1960-94 (in Japanese).

Mitsui, Shoji (1994). "Toshi sanson kankei kara miru ringyo rodoryoku no atarashii doko to igi" ("New trends of forestry labour power and their significance from the viewpoint of relation between cities and mountain villages"), *Ringyo Keizai Kenkyu (Journal of Forest Economics)* (125): 90-95 (in Japanese).

Miyazaki, Masayasu, and Ito, Osamu (1989). "Senji sengo no sangyo to kigyo." In Nakamura, T., ed., *Keikakuka to minshuka*, 165-235. Iwanami Shoten, Tokyo (in Japanese).

Nagata, Shin, and Terashita, Taro (1991). "Ringyo rodoryoku no yosoku ni tsuite no ichi shiron" ("An essay on forecasting labour force in forestry"), *J. Jpn. For. Soc.* 73(1): 50-53 (in Japanese).

Nonomura, Yutaka (1983). "Ringyo rodoryoku wo meguru shomondai." In Tsutsui, M., ed., *Rinseigaku*, 103-43. Chikyusha, Tokyo (in Japanese).

Okamori, Akinori (1979). "Shokibo ringyo keiei no hatten joken" ("Conditions for development of small-scale silvicultural management"), *Ringyo Keizai (Forest Economy)* (363): 9-17 (in Japanese).

Prime Minister's Office (1996). "Shinrin ringyo ni kansuru yoronchosa," *Rinyajiho* 43(3): 1-27 (in Japanese).

University of Tokyo Social Science Institute, ed. (1954). *Ringyo keiei to ringyo rodo*. 290 pp. Norin Tokei Kyokai, Tokyo (in Japanese).

Tanaka, Jun-ichi (1981). "Ringyo rodoryoku no jukyu yosoku," *Rinsei Soken Report* No. 15. 63 pp. Rinsei Soken, Tokyo (in Japanese).

– (1992). "Ringyo no rodo joken," *Rinsei Soken Report* No. 42. 53 pp. Rinsei Soken, Tokyo (in Japanese).

– (1994). "Ringyo rodo heno shinkisan-nyu," *Rinsei Soken Report* No. 45. 93 pp. Rinsei Soken, Tokyo (in Japanese).

6
National Forest Management
Koji Matsushita

Japanese national forest management started at the beginning of the Meiji era (1868-1912). Since then, national forests have been managed by the Ministry of Agriculture and Forestry, the Imperial Household, and the Ministry of Home Affairs, which managed national forests in Hokkaido. After the Second World War, the Ministry of Home Affairs was dissolved, and the forests managed by the Imperial Household were transferred to the National Treasury as payment of property taxes on Imperial Household holdings. On 1 April 1947, the three prewar national forest management systems were consolidated into one national forest sector, managed by the Bureau of Forestry, under the Ministry of Agriculture and Forestry. The total area involved was approximately 20% of Japan's total land area and approximately 30% of its total forested land area.

The Bureau of Forestry rose in status to become the Forestry Agency in 1947. A self-supporting accounting system was introduced (National Forest Special Accounting Law; Law No. 38 of 1947), beginning a new phase of national forest management. National forest management has faced a number of difficult financial problems since the mid-1970s, particularly the large cumulative debt, and the system was reformed in fiscal year (FY) 1998. In this chapter, postwar national forest management by the Forestry Agency will be discussed, with a particular focus on timber production and finance. In Japan, several ministries and agencies other than the Forestry Agency also own forestland, but because the total area held is small (197,000 ha, or 0.8% of Japan's total forestland), only the national forests managed by the Forestry Agency are discussed here.

National Forest Overview
At the end of FY 1994, the area of national forest was 7,647,000 ha, 30.4% of Japan's total forest area of 25,146,000 ha. Recently, the area of forestland managed by the Forestry Agency has been gradually decreasing because of

the sale of national forest land under the National Forest Management Reform Plan. The total national forest growing stock is 892 million m³, which is 25.6% of Japan's total growing stock. The conifer growing stock, 470 million m³, exceeds that of broad-leaved trees (423 million m³) because of the postwar expansion policy of planting conifers and the increased age of standing postwar plantations.

Both the Forest Law (Law No. 249 of 1951) and the National Forest Administration Regulation (Instruction of the Ministry of Agriculture, Forestry, and Fisheries, No. 21 of 1991; hereafter, NFAR) were amended in 1991. New forestland zoning methods were introduced in the 1991 NFAR, and the former 1958 NFAR zoning methods were abolished. Under the new regulations, national forest lands are divided into four categories (Forestry Agency 1990):

- *Land Conservation Forests,* which have as their priority functions prevention of disaster to the local community, public roads, water conservation, and maintenance of living environments
- *Nature Preservation Forests,* which give priority to ecosystem conservation, the protection of rare animals and plants, and the protection of the forest as a symbol of regional natural culture
- *Multiple Utilization Forests,* which focus on forest recreation, including resort facilities and the provision of nature education
- *Timber Production Forests,* which emphasize productive activities, including the harvesting of timber and the production of minor by-products.

The national forests are to be zoned into these four categories in the order specified. Timber Production Forests are specified last, as they consist of any remaining areas that are not classified under the other three categories. Water resource conservation functions are required for all categories. The areas of national forests classified according to these four categories are shown in Table 6.1. In 1996 the Timber Production Forest area was more than half of the total national forest area (54.3%). Land Conservation, Nature Preservation, and Multiple Utilization Forest areas were 18.8%, 18.5%, and 8.3%, respectively.

In the three non-Timber Production Forest categories, the percentage of natural forest is large. In Timber Production Forests, however, the areas of natural and artificial forest are comparable. In the 1991 NFAR, management practices for the new functions in each forest category were specified. In Land Conservation Forests, management practices include converting single-storied forests to multi-storied forests, planting trees for effective noise reduction and windbreaks, erosion control projects, improvement of forest road systems, and forest patrols. In Nature Preservation Forests, monitoring

Table 6.1

National forest resources in Japan classified by type of forest before the 1999 reform.

Type of forest	Total (Other + Forestland)	Other	Total (Forestland)	Forestland Timberland Artificial forest	Forestland Timberland Natural forest	Forestland Non-timberland
Land Conservation Forest	1,430	111	1,319	222	1,093	4
Nature Preservation Forest	1,414	351	1,063	14	1,049	1
Multiple Utilization Forest	636	60	575	137	437	2
Timber Production Forest	4,119	139	3,980	1,937	2,023	19
Outside of 4 categories	8	1	7	0	7	–
Total	7,607	662	6,945	2,311	4,609	25

Source: Forestry Agency (1998), Kokuyurinya jigyo tokeisho (50), 6-7.
Note: Areas are in thousands of hectares. Figures are from the forest operation plan (segyo kanri keikakusho), effective 1 April 1998.

the forest ecosystem, maintaining existing plant communities, preventing human damage, promoting educational activities, and setting up forest patrols are practices to be implemented. In Multiple Utilization Forests, construction of facilities for the preservation of health and culture, and enhancement of various types of forest to increase the complexity of the landscape and improve amenities, are to be implemented.

By maximizing the expected functions under the zoning requirements, the national forest management policy fundamentally promotes the maintenance of profitability. Specific policy objectives include: (1) land conservation, (2) conservation of water yields, (3) maintenance and creation of natural environments, (4) promotion of health and cultural preservation, (5) timber production utilizing various species, (6) forest resource improvement in non-national forests and fostering of the relationship between national and non-national forestry management, (7) improvement, guidance, and extension of forestry techniques, and (8) contribution to regional economic development.

Forestland zoning methods introduced in the 1991 NFAR were changed in the 1998 national forest reform, which abolished the NFAR and introduced the National Forest Basic Management Plan based on Article 4 of the National Forest Management Law (to be explained later). This plan included the basic management policy that recommended forestland zoning. In this new type of classification, introduced in the 1996 Basic Plan on Forest Resources, which was based in turn on the Basic Forestry Law, national forest lands are divided into three categories, the approximate percentages of which were determined in the National Forest Basic Management Plan: (1) Water and Soil Conservation Forests, 50% of the total area; (2) Symbiosis Forests, 30%; and (3) Sustainable Utilization Forests, 20%. The percentage of forest-related non-timber functions, namely, Water and Soil Conservation Forests and Symbiosis Forests, increased to 80%, from 50% in the 1991 classification.

The relationship between the 1991 and 1998 classification systems is as follows: Nature Preservation Forests and Multiple Utilization Forests in the 1991 classification correspond to Symbiosis Forests in the 1998 classification; Land Conservation Forests in the 1991 classification correspond to Water and Soil Conservation Forests in the 1998 classification; and Timber Production Forests in the 1991 classification are divided into Soil and Water Conservation Forests and Sustainable Utilization Forests in the 1998 classification. Soil and Water Conservation Forests are divided into two types: Land Conservation Type and Water Conservation Type Forests. Basic management policies in Soil and Water Conservation Forests include artificial forest management with a final cutting age that is older than that usually associated with artificial forests, and the creation of multi-storied forests, that is, a forestry management practice based on small, subcompartments

Table 6.2

National forest resources in Japan classified by type of forest after the 1999 reform.

Type of forest	Total (Other + Forestland)	Other	Total (Forestland)	Forestland Timberland — Artificial forest	Forestland Timberland — Natural forest	Forestland — Non-timberland
Water and Soil Conservation Forest						
Land Conservation Type	1,335	105	1,230	200	1,028	3
Water Conservation Type	2,780	82	2,698	1,217	1,471	10
Symbiosis Forest						
Nature Preservation Type	1,422	350	1,071	14	1,056	1
Multiple-Use Type	636	60	577	137	439	1
Sustainable Utilization Forest	1,415	62	1,353	731	615	7
Outside of 3 categories	8	1	7	0	7	–
Total	7,597	661	6,936	2,300	4,615	21

Source: Forestry Agency (2001), *Ringyo tokei yoran.*
Note: Areas are in thousands of hectares. Figures are from the forest operation plan (*kokuyurinya segyo jisshi keikakusyo*), effective 1 April 2000.

of forest areas set out in a mosaic pattern. Symbiosis Forests are divided into two types that are managed separately: the Nature Preservation Type Forest and the Multiple Use Type Forest. In the Sustainable Utilization Forests, forest practices are to be implemented that ensure that sufficient timber can be grown to meet diversification of timber demand while also considering forest health and non-timber utilities.

The total area of forest classified by the new zoning is 7.6 million ha, consisting of 4.1 million ha of Water and Soil Conservation Forests (Land Conservation Type, 1.33 million ha; Water Conservation Type, 2.77 million ha); 2.05 million ha of Symbiosis Forests (Nature Preservation Type, 1.42 million ha; Multiple Use Type, 0.64 million ha); and 1.44 million ha of Sustainable Utilization Forests (Forestry Agency 1999c). This new classification was introduced in the 1998 national forest reform, and the creation of forest operation plans and implementation of actual forest practices based on the new classification have just begun.

Regarding the contribution of national forests to regional economic development, three types of national forest use by local residents are permitted:

• The first relates to a profit-sharing reforestation system in which municipalities and private persons or groups can plant trees in specific national forest areas, with the final profit being shared between the Forestry Agency and the tree planters. As of 31 March 1998, this project had been allocated approximately 21,700 sites with a total area of 133,000 ha (Forestry Agency 1999a).
• The second type of use allows local residents to gather edible wild plants, mushrooms, and fuelwood for personal use. The total project area consists of about 1,700 sites covering 1,606,000 ha.
• The last type of land use involves leasing out forestland for construction of schools, public roads, dams, and so on. The total project area consists of approximately 57,100 sites on 80,000 ha.

The total area for these three types of contribution to regional economies is approximately one-fourth of the total national forest area. Although now declining because of national forest organization reform, businesses in the national forest also provide work opportunities for local residents.

Since Japanese national forests are generally located in mountainous regions, there are a large number of areas with high non-timber value. As of 1 April 1999, 63% of all national forests were legally allocated to protective areas, thus limiting forest practice to some extent (Forestry Agency 1999b). The legally specified national forest areas, and these areas as percentages of the total national forest, are as follows: Protection Forest *hoanrin* (Forest Law), 53% of the total area; Designated Area for Erosion Control (Erosion Control Law; Law No. 29 of 1897), 1%; Special Areas of Natural Park (Natural Parks

Law; Law No. 161 of 1957), 22%; Special Areas of Wildlife Protection (Law on Wildlife Protection and Hunting; Law No. 32 of 1918), 2%; and Historic, Scenic, and Natural Monument Forests (Cultural Properties Protection Law; Law No. 214 of 1950), 2% (the total exceeds 63% because of duplicate specifications) (Forestry Agency 1997a).

Forest practices that are restricted in these forests, including harvesting, are specified in the laws cited. In addition, the Forestry Agency is preparing protected forest (*hogorin*) in the national forest. *Hogorin* was first introduced in 1915 (Notification of Bureau of Forests, *Rin*, No. 144) and amended in 1989 (Notification of Director General of Forestry Agency, *Rinyakai*, No. 25). The 1989 Amendment divided the protected forest (*hogorin*) into the following categories (the figures in parentheses are the number of sites and their total area as of 1 April 2000): (1) Forest Ecosystem Protection Areas (26 sites, 320,000 ha); (2) Forest Biology Genetic Resource Preservation Forests (10 sites, 29,000 ha); (3) Timber Genetic Resource Preservation Forests (331 sites, 9,000 ha); (4) Plant Community Protection Forests (354 sites, 119,000 ha); (5) Specific Animal Habitat Protection Forests (31 sites, 16,000 ha); (6) Specific Geographical Protection Forests (33 sites, 30,000 ha); and (7) Forests of Special Significance to Regional Residents (32 sites, 2,000 ha) – a total of 817 sites and 526,000 ha (Forestry Agency 2001).

The most important areas are the Forest Ecosystem Protection Areas (FEPA). In the FEPA, old-growth or similar forests are protected, with the specific objectives of maintaining the natural forest ecosystem, protecting animals and plants, preserving genetic resources, developing forest practice and control techniques, and conducting scientific research. Each FEPA must include at least 1,000 ha of natural forests (old-growth forests or similar) that are representative of major Japanese forest stands. In limited areas of Japan, where rare stands are found, the minimum area is reduced to 500 ha.

The FEPAs are divided into Core Areas and Buffer Areas. Within Core Areas, the forest ecosystem is strictly protected. Buffer Areas are located surrounding the Core Areas. There, large-scale development is not acceptable but educational and recreational uses are permitted. As of 1 April 2000, there were 26 FEPAs totalling 320,000 ha. The largest is the Central Area of the Hidaka Mountains (66,353 ha) in Hokkaido. The smallest is situated on the east coast of Haha-jima in the Ogasawara Islands (503 ha). Yakushima Island (15,185 ha) and the Shirakami Mountains (16,971 ha), which were registered in 1993 with the Conservation for the Protection of World Culture and Natural Heritage program, are included as FEPAs.

Recently, the demand that forests have recreational functions has increased, and the Forestry Agency has attempted to increase the National Forest Recreational Area. As of 1 April 2000, there were 1,267 recreational areas covering 410,000 ha. The National Forest Recreational Areas include:

Nature Recreation Forests (91 sites, 105,000 ha), Nature Education Forests (171 sites, 35,000 ha), Scenic Beauty Forests (571 sites, 186,000 ha), Forest Areas for Sports (74 sites, 10,000 ha), Outdoor Sports Areas (241 sites, 53,000 ha), and Forests with Trails (119 sites, 21,000 ha). In FY 1999, the total number of visitors to these recreational areas was estimated to be 162 million (Forestry Agency 2001).

Timber Production

Figure 6.1 shows annual cutting and growth volumes, and Figure 6.2 shows the cutting and planting area. The volume cut increased annually from FY 1950 to FY 1964, peaking at 23,245,000 m³, and subsequently declined to 3,920,000 m³ (16.9% of the 1964 peak) in FY 1999. The postwar period can be divided into the following periods, delineated by fiscal years: Period 1 (1947-50), Period 2 (1951-59), Period 3 (1960-72), Period 4 (1973-77) and Period 5 (1978-98). As shown in Figure 6.1, Period 1 was the production-reduction period immediately following the Second World War. In Period 2, cutting activities rapidly recovered. Period 3 opened in FY 1960 with 20 million m³ cut annually. This production level generally continued until FY 1972. By 1973, however, the volume cut annually had begun to decrease rapidly, holding steady at about 15 million m³ during the 1970s and at the start of Period 4. The national forests' financial situation rapidly worsened

Figure 6.1

Cutting volume and annual growth in national forests in Japan, 1947-99.

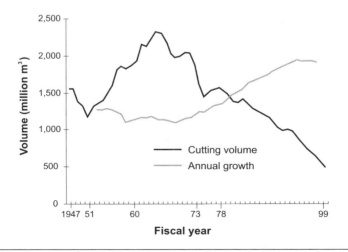

Sources: Forestry Agency (1946-96); Forestry Agency, *Kokuyurinya jigyo tokeisyo*, No. 50 (annual growth, 1997); Forestry Agency, *Ringyo hakusyo*, FY 1998-2000 ed. (cutting volume, 1997-99).

Figure 6.2

Area of cutting and planting in national forests in Japan, 1947-99.

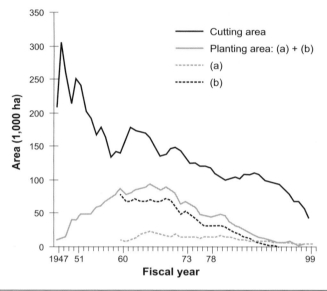

Note: (a) Reforestration ; (b) Expansion of artificial forest.
Source: Forestry Agency (1964), *Ringyo tokei yoran ruinenban;* Forestry Agency (1972, 1982, 1992), *Ringyo tokei yoran jikeiretsu ban;* Forestry Agency (1993-2001), *Ringyo tokei yoran.*

in the 1970s, and the start of Period 5 was marked by the new law instituting the National Forest Management Reform Plan.

Period 1 (FY 1947-50)
Period 1 immediately followed the 1947 consolidation, when, under a self-supporting accounting system, National Forest Special Accounting was introduced. The NFAR was established in 1948. It was the most important source of regulations relating to national forest management, including the formation of management practice rules. Thus began postwar national forest organization.

The NFAR introduced the following principles to national forest management (Shioya 1978):

- The management plan gained increased significance.
- The final cutting age, formerly that which maximized net profit, was changed to that which maximized mean annual growth or yield volume.
- The basis of yield regulation was changed from regulation of area to regulation of volume.

- The forest management unit for sustained yield was a working group; when sustained yield was impossible, the use of multiple working groups or larger working units was acceptable.
- The existing plan's 10-year planning period was replaced by a three-year working plan.
- The relationship between forest management and accounting, including the preparation of financial statements, was given importance in accordance with the introduction of the self-supporting accounting system.
- A cross-check process between management plans and actual practices was introduced.

Period 1 saw the introduction of postwar national forest management and various other new systems. The volume cut annually had been decreasing since 1947, as shown in Figure 6.1, but the area cut in FY 1948, about 300,000 ha, was the postwar maximum. Destructive cutting had begun during the war and continued afterwards. During FY 1947-51 the debt and financial problems were severe. Later, the NFAR was amended according to national forest management changes.

Period 2 (FY 1951-59)
During Period 2 the Japanese economy boomed and timber demand increased. Imported timber made up a small percentage of total domestic timber consumption. During this period, the Forestry Agency started two important timber production plans, the Long-Term Production Plan for the National Forest in 1955 and the Productivity Growth Plan for the National Forest in 1958.

Five major policies were introduced in the Long-Term Production Plan (Shioya 1978):

- Where artificial forests were acceptable, clear-cutting was practised and followed by the planting of marketable conifer species.
- Selective cutting was introduced where clear-cutting was unacceptable (e.g., protected or scenic beauty enhancement areas), in forests where natural regeneration was expected, in inaccessible forests, and in forests where artificial reforestation was impossible.
- In overmature old-growth stands, adjustment periods had to be prepared and the overmature forest had to be regenerated to marketable conifers as quickly as possible.
- In industrial timber production, the production of "usual-sized" logs (approximately 30 cm in diameter) had priority, and the locations of forests with "large-sized" logs (at least 50 cm in diameter) were to be limited.
- Fuelwood forests were to be reduced and their function changed to industrial wood production.

The idea of increasing national forest timber production and the concept of operating national forests under a unified management policy were introduced with the Long-Term Production Plan. As a result, each existing working unit's identification of forest resources and practices disappeared (Oka 1990). The Productivity Growth Plan (the management policy to increase timber production) offered a clearer picture, and established the following goals: (1) increasing artificial forests in the hope of doubling national forest productivity between FY 1958 and FY 1997, (2) reducing the final cutting age by using forest tree breeding techniques and forest fertilization, and (3) increasing forest road investment.

The NFAR, amended in 1958 by the introduction of the Long-Term Production Plan and the Productivity Growth Plan, was altered to permit an increase in national forest timber production. The 1958 NFAR introduced a zoning system for national forest management. Forest practices were restricted in relation to forestland non-timber values, as categorized in the Type 1 land category, and forest production appropriate to regional residents was required in the Type 3 land category. All forestland other than Types 1 and 3 were considered Type 2, where drastic forest management practice was possible (Oka 1990).

In addition, the management unit was expanded and the sustained yield working unit was abolished. The management plan area, exceeding the former working unit, was newly determined for the sustained yield unit, making the sustained yield volume easy to realize in the planning process. To determine allowable annual cutting volume for management plan areas, a new concept called Standard Cutting Volume, was introduced in the 1958 NFAR. Although Standard Cutting Volume was based on annual growth, it was also regulated such that annual cutting volume could exceed annual growth for the purpose of improving species and the type of forest cover when there were no problems in realizing a sustained yield. The three-year base of the 1948 NFAR working plan was replaced by an annual working plan that was to be revised based on actual practices. The expansion of the sustained yield unit and the adoption of a Standard Cutting Volume that exceeded annual growth allowed cutting volumes to be increased in national forests during Periods 2 and 3.

Period 3 (FY 1960-72)

Period 3, the height of prosperity for national forest management, saw annual timber production reach 20 million m³. When rapidly increasing timber prices became a political issue in 1961, the Ministry of Agriculture and Forestry introduced emergency measures to stabilize timber prices, including the new National Forest Timber Production Increase Plan of 1961 introduced by the Forestry Agency. Increased timber production, an objective of the new plan, was to be achieved by: (1) increasing the tree planting area,

(2) improving early-rotation tree growth by using forest fertilization, forest tree breeding, and improved tending practices, (3) reducing forest road construction, and (4) promoting mechanization.

In 1969 the NFAR was drastically amended to establish a production system corresponding to the increase in domestic timber demand. Standard Cutting Volume had to be calculated to include the annual growth after improvement practices. It was thus based on annual growth projections, calculated only by a desk plan and realized by the introduction of dense planting, shortened rotation, and the introduction of breeding (but without sufficient scientific and technical input). Considerations for recreational uses of national forests were also introduced in the 1969 NFAR.

Each new production plan in the 1950s and 1960s increased the number of both regular and non-regular national forest staff. The number of regular staff increased in Periods 2 and 3 as a result of the transfer of non-regular staff into regular staff positions. This increase led to harvest volume increases (Shioya 1978). Revenues and expenditures were well balanced because of the high timber prices and large cutting volumes. Subsequently in Periods 2 and 3, when the revenue and expenditure balance showed a loss in a particular year, a reserve fund was used to maintain positive accounts. Part of the profit was transferred to the national general account during FY 1959-72. The 20 million m^3 cut annually during this period was clearly overcutting, as it exceeded both the actual annual growth and the Standard Cutting Volume based on the future growth projected by the management plan. Such overcutting and increased staffing affected later national forest management. The annual cutting volume peaked in FY 1964, and annual growth reached its minimum in FY 1968. It was in the mid- and later 1960s that the cutting volume deviated most from growth. The cutting volume-to-growth ratio peaked at 2.19 in FY 1964. The characteristics of the planting activities of Period 3 (see Figure 6.2) show that the increased percentage of artificial forest (broad-leaved forest converted to coniferous forest) constituted a large part of the total area reforested, and that the proportion of reforestation (maintenance of coniferous forest) was still low.

Period 4 (FY 1973-77)

Period 4 is regarded as a turning point. This was when the problems in national forest management became clear, annual cutting volumes decreased rapidly, and the "New Forest Operation" in national forest management was introduced. It was also during this period that the Japanese economy shifted from rapid to slower growth as a result of the 1973 oil crisis.

In 1972 the Forestry Administration Council submitted a report identifying the following roles for national forest management: (1) the pursuit of public benefits from non-timber values, (2) planned and continuous supply of forest products, and (3) contribution to regional economic development.

The introduction of the New Forest Operation to realize these basic objectives redirected national forestry management policy to focus on: (1) decreasing cutting block area (maximum of 5 ha for protection forests [*hoanrin*] and 20 ha for other forests), decentralizing the cutting block, and, when clear-cutting, increasing the size of uncut buffer strips; (2) expanding appropriate natural forest operations in subalpine zones; and (3) increasing the number of protected forests (*hogorin*) to promote nature conservation and recreational use (Shioya 1978). These policy changes were stimulated by the overcutting in Period 3 and by increased social demand for nature conservation in national forests. Subsequent declines in timber production from 19,034,000 m³ (FY 1972) to 16,207,000 m³ (FY 1973) were unavoidable. From FY 1973 to FY 1977, production remained constant at approximately 15 million m³.

The New Forest Operation caused both timber production and timber productivity to decline. With the post-clear-cutting planting activities in Period 4 (shown in Figure 6.2), expansion of artificial forest decreased, and the reforestation area remained similar to that in Period 3. Staff numbers remained fairly constant despite declining timber production, rapidly worsening the financial situation. The annual revenue and expenditure balance was negative after FY 1970, except for FY 1973 and 1974. By FY 1977, the profit accumulated previously had decreased to ¥3 billion, and the last transfer to the national general account was made in FY 1977. In FY 1978, the accumulated profit was zero and the debt period began.

Period 5 (FY 1978-98)
During Period 5, when the volume cut annually was decreasing and debt was increasing, the financial situation worsened and the National Forest Management Reform Plans (discussed in the next section) were revised repeatedly. Both annual timber production and staff numbers declined. Production activities led directly to higher debt because of low timber prices and low productivity, thus decreasing timber production.

To maintain the income and expenditure balance, income other than that from forest product sales (for example, the sale of forestland and estates) was increased, peaking at ¥117 million in FY 1986. Income from the sale of forest products was ¥177 million in the same year; income other than from forest products was negligible. Since FY 1987, annual sales have generally declined, exceeding ¥50 million only in FY 1991. The total income from profit-sharing reforestation, forest recreational facility user fees, and land rents has slightly exceeded ¥10 million annually since FY 1985. Timber production income has decreased due to the declining trend in annual cutting volume and timber prices, which have stabilized at low levels. National forest log-sale prices average approximately ¥30,000-40,000/m³ (for example, ¥30,400/m³ in FY 1978 and ¥33,200/m³ in FY 1995).

Table 6.3

Changes in cutting area and volume according to cutting method in national forests in Japan, 1964-97.

Fiscal year	Method of cutting	Area (ha)	Volume (1,000 m³)			Volume (%)[1]		
			Total[2]	Coniferous forest	Broad-leaved forest	Total[2]	Coniferous forest	Broad-leaved forest
1964	Clear-cutting	110,689	19,197	9,950	9,248	83.1	43.1	40.1
	Selective cutting	50,174	2,738	1,748	990	11.9	7.6	4.3
	Thinning	77,663	1,156	876	280	5.0	3.8	1.2
	Total	238,526	23,091	12,574	10,518	100.0	54.5	45.6
1973	Clear-cutting	63,617	12,165	6,186	5,979	75.5	38.4	37.1
	Shelterwood cutting	2,957	325	214	111	2.0	1.3	0.7
	Selective cutting	63,975	2,919	1,761	1,158	18.1	10.9	7.2
	Thinning	24,620	699	580	119	4.3	3.6	0.7
	Total	155,169	16,108	8,741	7,367	100.0	54.3	45.7
1997	Clear-cutting	6,376	1,756	1,495	262	31.6	26.9	4.7
	Shelterwood cutting	1,046	126	79	48	2.3	1.4	0.9
	Selective cutting	48,690	1,673	978	695	30.1	17.6	12.5
	Multi-storied forest operation	634	81	79	2	1.5	1.4	0.0
	Thinning	40,884	1,916	1,778	138	34.5	32.0	2.5
	Total	97,630	5,553	4,409	1,144	100.0	79.4	20.6

1 Percentage of total volume of the year.
2 Subtotal does not agree with the total because figures have been rounded off.
Sources: Forestry Agency (1965, 1974, 1998), *Kokuyurinya jigyo tokeisho*, nos. 17, 26, 50.

For post-clear-cutting planting methods, the reforestation area finally surpassed the area of artificial forest expansion in FY 1987. This means that
some of the trees planted after the Second World War reached their final
cutting period and the area planted to conifers, which created uniform,
even-aged forests following the clear-cutting of broad-leaved trees, showed
a proportional decrease.

Changes in national forest cutting area and volume by cutting method
are presented in Table 6.3 for three periods: (1) FY 1964, when the maximum annual volume was cut; (2) FY 1973, when the New Forest Operation
was introduced; and (3) FY 1997, the most recently available data. Selective
cutting comprised 30.1% of the total volume in FY 1997, with clear-cutting
and thinning at 31.6% and 34.5%, respectively. The volume cut by methods other than clear-cutting, including pre-regeneration cutting, and the
volume produced from the silvicultural process of multi-storied forest operations, exceeded that of clear-cutting. In FY 1997, clear-cutting accounted
for 31.6% of the total cutting volume, clearly less than in FY 1964 and FY
1973 (83.1% and 75.5%, respectively). In contrast, selective cutting has been
increasing. The significant decline in clear-cutting, especially in the broad-
leaved trees, resulted from policy changes in favour of reducing the number
of new artificial forests created by clear-cutting natural forests. Since Period
4, the decline in the volume cut annually and the increased percentage cut
without using clear-cutting have decreased profitability.

The percentage from thinning has also been increasing, reaching 34.5%
in FY 1997 from 4.3% in FY 1973, because of the greater average age of the
artificial forests that were planted after the war. The increased need for thinning also affected profits, and thinning practices were not fully implemented
because of financial difficulties.

Another problem, related to natural regeneration, has arisen since the
introduction of the New Forest Operation in 1973. Controversy surrounds
the prewar national forest management practice of natural regeneration.
The 1929 Ministry of Agriculture and Forestry's Bureau of Forests budget
promoted intensive management in the national forests, including the practice of natural regeneration. The major points of controversy are the profitability in natural regeneration cost performance, intensive forest
management by forestry technicians, and the reduction of a management
unit's area (Mori 1983). The prewar introduction of natural regeneration
stimulated an increase in the budget for the Bureau of Forests in FY 1929
and in the number of district forestry offices. Post-harvest and silvicultural
cost savings were, however, due mostly to the recent introduction of natural regeneration after 1973. This introduction resulted in greater numbers of
forest stands where almost nothing was done after harvest because the number
of working staff and managers was declining due to financial difficulties.

Financial Situation

The Financial Situation and the Reform Plan

National forest management debt exceeded ¥3,800 billion by October 1998. This debt did not suddenly appear, having accumulated over a long period (Table 6.4). In FY 1978, the income and long-term debt were ¥256.8 billion and ¥99.7 billion, respectively, and nearly all expenditures were for business – ¥363.8 billion for business and ¥10.5 billion for bank interest. At that time, however, when ¥226.8 billion was the income from forest product sales, personnel expenses amounted to ¥251.3 billion, almost equalling self-income of ¥256.8 billion (self-income is all income except General Accounting aid and debt, including income from sales of forest products, forest by-products, and forestland; recreation facility entrance fees; and income from other national forest activities). Thus, income other than from forest product sales was promoted, for example, sales of forestland and estates, which amounted to ¥17.7 billion. The balance between national forest revenue and expenditure was in deficit during FY 1970-72, despite an exceptional ¥95.9 billion annual profit in FY 1973. This annual profit was at a maximum because of the rapid price increases caused by the oil crisis, and the financial problem that was already serious remained hidden.

Total income, including debt, increased until FY 1993 and has decreased since then. Self-income peaked at ¥325.8 billion in FY 1979 and decreased to ¥96 billion in FY 1998. Long-term debt has shown a tendency to increase, finally exceeding self-income in FY 1991. Financial help from the national general account has been increasing gradually, thus decreasing the ratio of self-income to total income to 20.8% in FY 1998. In that fiscal year, product sales of ¥44.6 billion provided about half of the total self-income. The other half consisted of income from sales of forestland and estates (¥27.4 billion), miscellaneous income (¥9.9 billion), financial aid from the national general account (¥52.5 billion), and a financial grant from the erosion control account (¥14 billion). By FY 1998, the percentage of total income from forest product sales had decreased to 9.7%.

Total expenditure has been between ¥300 billion and ¥400 billion since FY 1976, peaking at ¥413.8 billion in 1981. The long-term debt interest and repayment load has gradually increased, approaching the level of business expenditures in FY 1995. Since FY 1992, total interest and repayments have exceeded income. In FY 1994, total interest and repayments exceeded total income and aid from the national general account, and since FY 1994 all income, including aid from the national general account, has been used to service long-term debt while the deficiency has been covered by new debt. Consequently, all business expenditures, including personnel expenses, are now also covered by new debt. The self-supporting accounting system introduced in 1947 has completely failed.

Table 6.4

Financial situation of national forests in Japan, 1976-98.

Fiscal year	Income				Expenditure				Cumulative loss	Balance of debt
	Total	Self-income[1]	Aid from General Accounting	Long-term debt	Total	Expenditure for business	Interest and repayment	Profit		
1976	3,042	2,642	–	400	3,244	3,230	14	-504	(973)[2]	400
1977	3,393	2,563	–	830	3,573	3,554	19	-906	(30)[2]	1,230
1978	3,613	2,568	48	997	3,743	3,638	105	-991	961	2,227
1979	4,518	3,258	80	1,180	3,875	3,704	171	-319	1,280	3,407
1980	4,486	3,062	84	1,340	4,221	3,955	266	-657	1,937	4,736
1981	3,976	2,489	87	1,400	4,560	4,138	422	-1,472	3,409	6,080
1982	4,488	2,701	87	1,700	4,600	4,000	600	-1,060	4,469	7,654
1983	4,827	2,665	92	2,070	4,778	3,970	808	-699	5,168	9,509
1984	4,977	2,609	98	2,270	5,084	4,030	1,054	-868	6,036	11,461
1985	5,031	2,605	106	2,320	5,125	3,819	1,306	-786	6,822	13,350
1986	5,652	3,167	115	2,370	5,266	3,703	1,563	-159	6,981	15,140
1987	5,570	2,881	131	2,558	5,518	3,739	1,779	-542	7,523	16,980
1988	5,757	2,907	150	2,700	5,676	3,733	1,943	-535	8,058	18,876
1989	5,839	2,962	177	2,700	5,690	3,627	2,063	-436	8,494	20,726
1990	5,564	2,730	194	2,640	5,769	3,610	2,159	-719	9,213	22,511
1991	5,571	2,314	269	2,988	5,888	3,580	2,308	-1,177	10,390	24,630

1992	5,675	2,333	363	2,979	5,805	3,375	2,430	-1,060	11,450	26,730
1993	6,263	2,244	511	3,508	5,927	3,378	2,549	-1,066	12,516	29,291
1994	5,619	1,915	568	3,136	5,929	3,270	2,659	-1,242	13,758	31,429
1995	5,322	1,780	573	2,969	5,675	2,842	2,833	-1,318	15,076	33,308
1996	5,482	1,767	569	3,145	5,555	2,549	3,006	-1,067	16,143	35,228
1997	5,504	1,318	609	3,595	5,526	2,358	3,168	-1,395	17,538	37,446
1998	4,602	959	525	3,119	4,546	2,175	2,370	-1,008	18,546	–

Note: All amounts are in 100 millions of yen.
1 All income except General Accounting aid and debt. Includes erosion control aid.
2 Accumulated surplus.

Sources: Tabuchi, Yuichi (1997), "Zaimukara mita kokuyurin" ("National forest from financial aspects"), *Ringyo Keizai (Forest Economy)* (587):32; Forestry Agency (1999), *Kokuyurinya jigyo no bapponteki kaikaku,* Nihon Ringyo Chosakai, Tokyo, 415; Forestry Agency (1999), "Kokuyurinya jigyo tokubetsu kaikei kessan no gaiyo," *Rinya-jihou* (542):15 (for 1998 data).

National forest management has depended on external funding, especially long-term debt, since 1976. Alleviating such financial difficulties required drastically improving the management system itself, and the Special Measures Law of the National Forest Management Reform (Law No. 88, hereafter referred to as the Reform Law) was enacted in 1978. This law was intended to re-establish a balance between revenues and expenditures by FY 1997, while at the same time maintaining the self-supporting accounting system. The first National Forest Management Reform Plan (hereafter referred to as the Reform Plan), for planning period 1978-87, was introduced to balance finances by FY 1997. The first Reform Plan's major points were: (1) improvement of management practices, (2) achievement of reasonable staff numbers, and (3) maintenance of self-income.

Concrete measures regarding the second point involved first simplifying the national forest organization, including staff reductions, and then unifying or abolishing regional, district, and ranger station offices. The first Reform Plan failed to re-establish a financial balance, however, and subsequent Reform Law and Reform Plan amendments were more severe. The Reform Law was amended in 1984 (Law No. 36) to form the second Reform Plan (planning period 1984-93), under which it became possible to address financial problems, such as long-term debt due to staff retirement payments. The second Reform Plan also specified that the total number of regular and non-regular staff must be reduced to 40,000 by the end of FY 1988. By the third Reform Plan of 1987 (planning period 1987-96), the amended Reform Law (Law No. 77) specified a reduction of staff numbers to 20,000. The fourth Reform Plan (planning period 1991-2000) started in 1991, and the final year to balance finances was changed from 1997 to 2010.

Staff numbers have therefore declined considerably under these Reform Plans. As of 1 April 1998, they had declined from 65,000 (including 35,000 regular staff) at the start of the first Reform Plan in 1978 to 14,000 (including 8,000 regular staff). Furthermore, the number of district offices also declined, from 351 to 229, and the number of ranger stations declined from 2,333 to 1,256. The volume of trees cut annually declined from 15.3 million m³ in 1978 to 6.5 million m³ in 1997.

There are two types of national forest timber production systems: the Direct Management System, where national forest staff are directly involved with timber production, and the Contract System, where private companies work under contract. The productivity of the Contract System generally exceeds that of the Direct Management System, and the percentage contracted out increased from 24% in 1978 to 61% in 1995. Maintenance of income has also improved; sales of non-forest products, for example, have increased. Nevertheless, the recent financial situation is such that the total self-income and the aid from the national general account amounted to less than the total of interest and repayments. The necessity of taking on new long-term debt for interest and debt retirement has revealed that revenue and expenditure balance will not be achieved under the self-supporting accounting system.

In December 1996, the Japanese government Board of Audit reported the national forest management's financial failure. This necessitated a drastic reform of debt treatment. The Board of Audit declared that organizational simplification by the former Reform Plan was insufficient to decrease this debt, so new solutions were required. On 25 December 1996, the Cabinet decided on an administrative reform program that included national forest management reform.

First, it was recommended that a national forest management reform program be created. Reforms included a more drastic rationalization of management that would lead to the simplification and rationalization of the organization as well as a planned reduction of working staff. This management reform program was to be created in close cooperation with ministries and agencies related to the national forest system.

Second, it was decided that during 1997 laws relating to national forest management reform would be prepared for an ordinary session of the Diet of 1998, and that the financial measures necessary for the reform would be taken. The Forestry Administration Council announced a national forest management reform draft report in July 1997 that identified the following policy directions:

- changing national forest management objectives from timber production to land and environment conservation
- redirecting worksite operations to the private sector
- introducing the river-based forest management system and the creation of a fund supported by the population, municipalities, and enterprises in a river's lower reaches
- reconsideration of the self-supporting system and revenue sources
- reconsideration of measures that are new and independent of a reconsideration of the self-supporting system to reduce the cumulative debt.

The draft report delineates radical changes for national forests, including the abolition of the self-supporting system introduced in 1947. The Governmental Investments and Loans' Fund Operation Council, a lender to the national forest system, reported in August 1997 that no new loans would be made to the national forests. The management that had led to the accumulation of repeated debts had finally led to complete bankruptcy.

Change of Accounting System in 1998
The Forestry Administration Council announced a final report on national forest management reform on 18 December 1997. The report demonstrated

and emphasized that the basic idea for national forest management reform is that the national forests must be managed with the participation
of the people, for the people, as the common property of the people. Thus
it becomes a "Forest for People."

The report pointed out that the most important national forest management objectives must focus on non-timber utilities. These utilities are to be
related, in the main, to national land conservation and to environmental
functions for timber supply in line with the changes in social demand with
respect to the national forest. The final report also pointed out the necessity
of abolishing the self-supporting accounting system and of introducing a
new special accounting system that included support from General Accounting. These recommendations corresponded to the change of management
objectives.

On 25 December 1997, Cabinet decisions relating to a definite plan for
long-term debt repayment by the former Japanese National Railways and
national forest management reform were reached. Four points were determined for national forest management reform:

1   The main objective of national forest management would change from
    timber production to public utilities.
2   The organization of national forests would be rationalized and the number
    of workers reduced, after careful consideration of the employment situa-
    tion and the relationship between labour and management.
3   The accounting system would be changed from the self-supporting spe-
    cial accounting system to special accounting that would include support
    from General Accounting.
4   With respect to the long-term debt, the premise of self-directed action by
    national forest agencies was of fundamental importance. Nevertheless,
    given the existence of a debt level that the national forest could not
    repay, some definite measures, such as subrogation by General Account-
    ing, would be necessary.

In October 1998, the Special Measures Law on National Forest Reform
(Law No. 134) was enforced, and it was decided that the long-term debt of
approximately ¥3,800 billion should be divided into two categories. First,
¥2,800 billion of debt was moved to General Accounting from National
Forest Special Accounting in order to prevent further accumulation of debt.
The capital repayments on this debt were to be paid by General Accounting, while the interest was to be paid by revenue from the tobacco tax and
government bond expenditures. The other ¥1,000 billion still remaining in
National Forest Special Accounting remained as long-term debt; National
Forest Special Accounting had to repay the capital over the next 50 years
through profits produced by forest product sales and by sales of forestland

and estates. The interest on this debt was to be paid from General Accounting in order to prevent further accumulation of debt. In October 1998, the National Forest Special Accounting Law was therefore amended and the new accounting system was introduced. Although this new accounting system is also basically a special accounting system, expenditure related to national forests where the management objective is not timber production can be paid from General Accounting. In other words, the self-supporting accounting system introduced in 1947, when the current national forest system was created, was abolished.

Point 4 of the December 1997 Cabinet determination on national forest reform pointed out that the issues relating to long-term debt had to be considered on the basis of internal action by the Forestry Agency itself as far as possible. It was estimated that the Forestry Agency could account for ¥1,000 billion. With respect to the payment of this ¥1,000 billion, the Forestry Agency published a 50-year outlook report on national forest income and expenditure that shows the payment schedule for the amount over the next 50 years (Forestry Agency 1998b, appendix, 8-9).

This report is based on several key assumptions. Sales of forest products and an annual cutting volume that is strictly linked to timber sales are understood to be the most important factors relating to this income. The report assumes that annual cutting volume in the national forest, including thinning, will increase to 6.7 million m^3 in FY 2004-08, 8.4 million m^3 in FY 2009-13, 11.9 million m^3 in FY 2014-18, 13.2 million m^3 in FY 2019-23, 14.6 million m^3 in FY 2024-28, 15.3 million m^3 in Y 2029-33, 15.8 million m^3 in FY 2034-38, 15.8 million m^3 in FY 2039-43, and 15.9 million m^3 in FY 2044-48. The five-year annual cutting volume, however, actually decreased from 9.0 million m^3 in FY 1993 to 8.3 million m^3 in FY 1994, 7.6 million m^3 in FY 1995, 7.0 million m^3 in FY 1996, 6.5 million m^3 in FY 1997, 5.6 million m^3 in FY 1998, and 4.9 million m^3 in FY 1999.

The Forestry Agency's outlook report assumes that the annual cutting volume will continue to decrease in the near future, but that the trend will reverse later, when the artificial forest planted after the Second World War matures to its final cutting age. This cutting prospect is based on the current Basic Plan on Forest Resources (planning period 1 April 1997 to 31 March 2012), which was introduced in 1996 and was based on Article 10 of the Basic Forestry Law (Law No. 161 of 1964). It is worth noting that the Basic Plan shows only possible production, calculated from the age-class distribution of artificial forest resources, and does not consider market conditions.

Another important assumption in the outlook report is expenditure. The repayment calculations include the slice of management expenses saved by the reduction of staff. In this assumption, the number of working staff is reduced to approximately one-third of the 15,000 people who made up the numbers at the end of FY 1996. This reduction is based on point 2 of the

Cabinet determination of December 1997. Thus, the repayment plan, which aims to pay ¥1,000 billion over the next 50 years, strictly depends on a future increase in demand and reduction of the workforce. It appears that full repayment according to this plan will be difficult to realize.

Management Failure and Its Accounting System

Several of the problems that led to the financial failure of the national forest are commonly observed throughout Japanese forestry, including the private sector. In this section, however, the focus is the national forest's financial failure. Some comments on the introduction of the postwar national forest special accounting system are included here.

The special accounting system was appropriate for national forest management as a government establishment. It was difficult, however, to maintain government enterprises at business levels of efficiency without a suitable accounting system. Management failure was caused not by the special accounting system itself but by the following system applications: (1) the Growing Stock Account, a basic idea introduced to the special accounting system before 1972; (2) the self-supporting system, and (3) the introduction of, and dependence on, long-term debt with interest. These issues will be explained further in this section.

The national forests had been managed through a self-supporting system in accordance with the National Forest Special Accounting Law of 1947. Despite financial difficulties, this system had been maintained in a form where both debt and aid from the national general account were calculated as income. The national forest basic accounting method was changed in 1972, and has continued in this altered form since 1973. Before 1972, the Growing Stock Account was used as a basic accounting concept. This idea was based on the management of a normal forest. Maintaining the forest requires annually balancing cutting and regeneration areas. The idea was adapted from the Normal Stock Method from general accounting theory. Thus, in the forestry application, normal growing stock became relevant in addition to the normal stock. Since actual cutting and regeneration activities are not equal in such an ideal situation, the Cutting Adjustment Account and Reforestation Adjustment Account were prepared.

In the Cutting Adjustment Account, differences between the growth (used to determine cutting activities) and actual cutting volume are adjusted. In the Reforestation Adjustment Account, differences between the reforestation volume necessary to maintain growth and the actual reforestation volume are adjusted (Forestry Agency 1986). When the cutting plan is based on normal forest growth and the cutting area is to be reforested, a normal age-class distribution will be maintained. Here, reforestation cost on the cutover land is considered a cost for creating harvest income.

However, the Growing Stock Method, which is based on the normal forest concept, has several problems with real applications. Age-class distribution is not uniform in national forests, and annual cutting volumes during the 1960s were double annual growth. Since 1973, the Growing Stock Method has been changed to the type of accounting system used in general enterprises. One reason for this change was the necessity to clarify, from an accounting viewpoint, the anticipated regeneration process that was to be conducted with debt funding (Tabuchi 1997). The Growing Stock Method was ineffective in preventing deterioration of the forest resource. The accounting system considered only the balance between total cost, including silvicultural activity costs, and total harvesting revenue. Recognizing that national forests are not normal forests, when cutover land is reforested annually, total harvest revenue will exceed reforestation costs. Considering the failure to expand reforestation on national forest land, however, this basic inequality is also questionable. This idealistic method, which failed to take all the factors adequately into account, ended when overcutting was conducted in Periods 2 and 3. Accounting's general role is to clarify management results and the financial situation. However, introducing the special adjustment accounts covered up the large imbalances between growth and harvest volumes by implementing a system capable of ignoring differences between reality and the normal forest ideal.

To highlight the problems of the self-supporting system, we need to begin with the introduction of the special accounting system into national forest management. Arguments about the introduction of this system began before the Second World War, with discussion falling into three categories (Maki 1965): (1) rationalization of management and capital maintenance; (2) adjustment of supply, demand, and the price of forest products; and (3) national finance.

Rationalization of management and capital maintenance related to the profit-making objective, part of which included the national forest's contribution to national finance. This objective was included in national forest management goals at that time. Although national forests were required to be managed for public benefit, accurate profit and loss statements were necessary to promote rational management. Thus, the special accounting system was deemed necessary. Capital maintenance, especially maintaining growing stock as capital, was also considered important. Annual revenue and expenditure calculations often allowed for harvesting activities but ignored the maintenance of growing stock; the money thereby lost to national finance resulted in forest resources of lower quality.

National finance, the third category, was a counterargument to the prewar situation, when the national forests had been managed by the Ministry of Agriculture and Forestry's Bureau of Forests under the national general account. Satisfying the Ministry of Finance's demands for fixed payments meant that the volume cut increased when timber prices fell and decreased when they rose, a policy that did not contribute to forest products supply-and-demand adjustment or to price stabilization. As the largest supplier, the national forest system, with cutting activities regulated under a unified accounting system, should have been allowed to make adjustments based on timber supply and demand. Introducing a special national forest accounting system but adopting a cutting policy that reversed the prewar approach (i.e., decreased cutting volume when timber prices declined and increased it when they rose) would have brought a larger contribution to national finance. The prewar national forest system was managed under a national general account that kept the balance in the black. Although national forest officials wanted the special accounting system, it was unacceptable to the national general account because a valuable financial source was being lost.

The three points argued in prewar discussions are also pertinent when examining the special accounting system introduced after the war.

For the first point, rationalization of management and capital maintenance, cutting volume exceeded growth volume during Periods 2 and 3, with the result that all growing stock maintenance was almost completely ignored. Management rationalization had not advanced either. Although the postwar accounting system was supposed to be specialized, it was constrained by the requirement that revenue and expenditure balance annually and that part of the annual profit be transferred to the national general account. The new special accounting system did not consider the first point sufficiently. From this perspective then, it was not the system that the prewar government officials of the Bureau of Forests desired. In the sense that annual cutting volume increased during Periods 2 and 3, the second and third points were almost accomplished by the special accounting system. Recently, Japan's timber self-sufficiency rate decreased to approximately 20%. In addition, the diminished economic importance of the national forest as a source of national finance revenue and declining regional economies have diminished the role of the second and third points. Finally, ignorance of the first point during Periods 2 and 3 continues to influence current national forest management.

Since the beginning of postwar national forest management, including land conservation projects, the self-supporting system has caused problems. There is room for the prewar arguments about the special accounting system in national forest management, which consisted mainly of timber production. To correct the destructive cutting during and immediately after the war, the national land restoration project was implemented under the special accounting system. Such projects should be completed as public utilities under the national general account. The introduction of the special accounting system, covering costly land conservation projects, appeared to indicate that national forest management should not anticipate future profits. Nomura (1956) said that the Ministry of Finance introduced the special accounting system from a financial viewpoint. Mr. Tatsuo Miura, the first Director General of the Forestry Agency (1947-50), held that the objective of introducing the special accounting system was to increase the amount transferred to the national general account by increasing labour efficiency (Mori 1983, 45). In view of the national financial situation, Mr. Sakae Shibata, the third Director General of the Forestry Agency (1952-55), saw the danger in continued management of the national forests under the national general account. He indicated that the idea of consolidation and the introduction of the special accounting system was agreed to by all forestry technicians (Mori 1983, 105).

Despite several differing evaluations of the introduction of the special accounting system in 1947, by 1952 the high postwar timber price meant that no cumulative debt occurred immediately and transfers into the national general account began in 1953. An equilibrium was thus established between revenue and expenditure that included national forest land conservation projects. Until 1978, the first year of aid from the national general account, expenditures for land conservation projects that used subsidies or public works for private forests were handled by the special accounting system, where revenue was almost entirely limited to forest product sales. Since national forests generally were located near mountain ridges or in inaccessible areas, land conservation project expenditures were not ignored. The National Forest Special Account could not accumulate anual profits, however, because the national forests had to pay the costs of the land conservation project and transfer part of the annual profit to the national general account.

Although public benefit was an important aspect of the national forest's role, there was no allowance for public benefit in the accounting system, or for the roles of the national forest other than timber production, such as the creation of recreational facilities and wildlife management. In postwar national forest management, maintenance activities and the increased emphasis on non-timber values were considered only in terms of cost, with no consideration of revenue. Thus, introducing the self-supporting system meant cheap national forest management at a time when the national finance was itself in difficulty. After the 1970s, however, this rationale disappeared.

The introduction and continuation of long-term debt and its associated interest has become another serious problem. Since 1976, the national forest has incurred increasing long-term debt from government investments and loans, mainly from postal savings, accumulated employees' pension insurance funds, and national pensions. The long-term debt (approximately

¥3,800 billion as of 18 October 1998) divided according to the rate of annual interest is as follows: less than 4% (29.5%), 4-5% (26.1%), 5-6% (16.9%), 6-7% (15.1%), 7-8% (11.5%), and 8% and over (0.9%) (Forestry Agency 1999a). Since 1992, total interest and capital repayment has exceeded total self-income. The internal rate of return for reforesting sugi (Japanese cedar, *Cryptomeria japonica*) has been decreasing: 6.3% in 1965, 4.1% in 1975, 2.1% in 1985, and 0.9% in 1992 (Forestry Agency 1994). The main reasons for the decrease were a fall in timber prices and an increase in costs such as wages. Under such economic circumstances, the national forest continued borrowing from government investments and loans at high interest rates.

The economic difficulties facing both national and private forest management systems were both very real and understandable. For private forests, a long-term loan program was established with a low interest rate that took into consideration the number of years necessary for timber production and the public benefits that accrue from regeneration. For example, as of 24 September 1997, reforestation loans from the Agriculture, Forestry and Fisheries Finance Corporation used a 2.50-2.65% interest rate with a 30-35-year repayment period. The fact that national forests previously could not utilize such long-term, low-interest loans is one reason for the current increase in cumulative debt. When private companies with large debts could not pay the interest, before the application of the Company Resuscitation Law, they could request relief measures such as interest reduction and exemption. Because national forests were not allowed to go bankrupt, such measures were not available to them and they continued borrowing from government investments and loans. A similar situation was experienced with the former Japanese National Railways when it was managed by the government. The Japanese National Railways were privatized in 1987 into regionally based private railway companies.

Thus, a significant problem in the special accounting system forced national forests to transfer part of their profits to the national general account and prohibited them from reserving funds when management was in the black. Conversely, when in the red, they were forced to borrow with interest. Given that transfer to the national general account was required based on the annual revenue and expenditure balance, deficiencies occurring in any particular year were to be settled in that year. Similar financial deterioration, caused by easy borrowing, occurred in various parts of the Japanese public sector, and the financial failure of the national forest system is a typical case (Ouchi 1989). Despite the introduction of short-term measures to rectify the deterioration, such as the increase in income from forestland sales and decreased staff numbers, it became clear that the national forests could not repay their large cumulative debt. The long-term negative effects of continuing short-term measures on national forest resources could no longer be ignored.

Organizational Characteristics

Management Organization

The national forest system is divided into nine regional forestry offices (Sapporo, Aomori, Akita, Maebashi, Tokyo, Nagano, Osaka, Kochi, and Kumamoto) and five branch forestry offices (Asahikawa, Kitami, Obihiro, Hakodate, and Nagoya). This organization was completed by February 1999 (hereafter regional forestry offices are called "regional offices" and branch forestry offices are called "branch offices"). As of 1 April 1996, there were 264 district offices, 75 forestry management centres and forestry management offices, and 70 specific project offices (69 of which were for erosion control projects).

Because of organizational simplification and the large cumulative debt, the number of district offices has been decreasing. Forestry management centres and forestry management offices have been created from former district offices, and continue to conduct almost the same activities. The number of ranger stations, the lowest national forest unit, has also been decreasing. On 1 March 1999, nine regional offices and five branch offices were reorganized to create seven new regional offices, and 264 district offices were reorganized into 98 new river-based district offices (Forestry Agency 1999a).

Two characteristics relevant to national forest organization should be mentioned. First, the work of regional and district offices is strictly limited to national forests owned by the Forestry Agency. They have almost nothing to do with private forests or national forests owned by ministries and agencies other than the Forestry Agency. Since the forestry promotion program for private forest owners (with a high subsidy rate) did not exist before the Second World War, district offices worked for both national and, to some degree, private forestry management. For example, they provided private forestry management with technical guidance and assistance in making regional development plans (Mori 1983).

Besides owning and managing national forests, the Forestry Agency produces and implements general policy for national and non-national forest resources and forestry. The district offices administer national forests, and prefectural government forestry departments administer non-national forest lands. The two are completely separate. For example, until the 1991 Forest Law amendment, the forest planning system, the most important forest resource policy regulated by the Forest Law, was divided into national and non-national forest components. Under the 1991 Amendment, the Forestry Agency established a new, common regional forest planning area for both national and non-national forest plans, which were made with mutual consideration. There is, however, almost no relationship between these plans now.

Second, the low level of national forest timber production has become an issue. National forest staff numbers gradually increased along with increases in cutting volume in the 1950s and 1960s. As the postwar national forest was to be managed on a commercial basis, under a special accounting system, labour productivity had to be competitive with that of private companies. Despite special accounting system management, however, national forest staff were public officials with guaranteed salary levels. Thus, the volume of national forest timber production per worker-day was clearly below that of private companies, while national forest workers' wages exceeded those of their private counterparts.

Labour-management disputes became severe in the latter 1950s. The National Forestry Workers' Union of Japan was one of the country's strongest labour unions. The district office director's most important work as chief forester should have been the total management of the national forest within the jurisdiction. This focus was diluted when labour-management conflict required time-consuming negotiations. The antagonistic relationship between labour and management led to a low point in labour productivity. In some district offices, over half the annual working days were spent negotiating the conflict. In 1966 this led the Minister of Agriculture and Forestry to state in the House of Councilors that national forest management was based on stabilizing employment under the Direct Management System principle by national forest workers (Mori 1983).

Although this principle was to maintain forestry jobs, it made national forest management worse. Thus, the Reform Plans emphasized increased use of contractors. The result of the investigation into national forest labour productivity by the Administrative Inspection Bureau of the Management and Coordination Agency (1990) showed that labour under the contract system exceeded productivity levels achieved under the Direct Management System. This result took into account the fact that, in general, the operation site was a good location for the Direct Management System as compared with the Contract System. The Contract System's unit wage was almost half that of non-regular national forest workers. The low productivity of national forest workers and former labour-management conflicts together resulted in both deterioration of national forest management and a decreasing reliance on national forests by the private forestry sector. The private sector's distrust of the national forest appears to be connected to the fact that national and non-national forests are managed separately and administered without mutual consideration.

Problems with National Forest Management
In 1998 cumulative debt exceeded ¥3,800 billion and the national forest management system was reformed. There are, however, national forest

management problems other than financial difficulties. Four such problems are highlighted here.

The first problem concerns the Forestry Agency, which works with both national forest management and the general forest policy administration, and the public's inability to distinguish the Forest Agency's national forest management role from its other activities. The Forestry Agency has three departments: Forest Policy Planning, Private Forest, and National Forest (the latter resulted from the merger of two departments on 1 March 1999: National Forest General Affairs and National Forest Management). The first two departments relate mainly to private forests. In 1971 the Japanese Committee for Economic Development published a report ("Direction to a Green Plan in the 21st Century") that introduced a new concept of agency creation. The idea was that a new agency could be made by separating the departments related to national forest management from the existing Forestry Agency (Shioya 1978). The new Forest Agency's main task would be restricted to general forest policy only.

Since the national forest's finances were not facing severe cumulative debt at the time, a new plan was possible. Recently, however, the severe financial crisis has complicated the discussion of the national forest organization and of what is best for the Japanese forestry system as a whole. It seems that both successfully producing timber and making general forest policy is difficult for one administrative organization to achieve. This problem had been discussed in 1965, when the Central Forest Council reported that the national forest should be managed by a legally independent public corporation. Although the large debt needs reform, the debt itself constitutes an obstacle to drastic reform. Despite the large debt, the most difficult problem politically is how to transfer the national forest workers, who are now national public officials, to non-public positions with significantly lower salaries.

The second problem concerns the relationship between forest policy and land, or environmental, policy. The forest planning system under the Forest Law is not well integrated with land-use planning policy and environmental policy. The national forest system has generally been expected to deal with water reserves and nature protection, whereas forest policy has always been separate from land or environmental policy. For example, program administration and areas that are managed as Forest Ecosystem Protection Areas often overlap with areas of the national parks that are under the Environment Agency. Limiting the work of district offices to timber production and land conservation that consists mainly of erosion control projects is one problem. Another is that wildlife management is handled by the Environment Agency rather than the Forestry Agency, meaning that Forestry Agency district offices have few, if any, wildlife management specialists on their team, and protecting recently planted trees from wildlife damage has become a problem.

The linking of land and environmental administration must be considered both with regard to national forest reform and in relation to private forest policy. The 1991 National Forest Administration Regulations specify that all national forests play a role in water conservation and that the lowest national forest management unit should be realigned in accordance with that objective. This same perspective and approach can also be applied to national forest tourism and recreation. Japanese national parks are a zoned system where many national park areas overlap national forest land, but few district offices have special tourism or recreation staff. Some district office staff do focus on tourism and forest recreation facilities, but their work is limited to forest and in-forest facility management related to recreational use rather than direct contact with visitors. Nature guides for visitors have recently been placed in district offices. There are many National Forest Recreation Areas, the most important of which are the Nature Recreation Forests. Since district office staff do not deal directly with visitors, Nature Recreation Forest management is generally entrusted to outside management. Simplifying national forest organization while maintaining the former organization's structure, which is focused on timber production, complicates and hinders the addition of a new environmental programs section.

The third problem concerns the uniform management methods applied in national forests. National forests exist in different environmental conditions, making uniform management difficult and inappropriate. The plan to restore Japan from destructive wartime and postwar cutting made promoting management practices nationwide seem sensible. Under the tight supply-and-demand situation in the 1960s, promoting a nationwide increase in national forest cutting volumes also seemed sensible. This, however, amounted to forcing a uniform forest policy upon the national forest system. With the recent financial difficulties, cost reduction measures and increases in non-timber revenues have received nationwide emphasis.

Besides differing environmental conditions, there are other regional differences with regard to national forest expectations. Different management policies suited to different areas must be considered, at least at the regional office level. National forest management unification also appears to be related to the system's personnel management. Currently the tenure of district office directors and forest rangers is only two to three years, and occasionally as little as one year; there is no provision for long-term tenure (for example, 10-20 years). Since the current district office's management area is not a sustained yield unit, concern for stability in the position of district office director has diminished (Mr. Masahide Katayama, the ninth Director General of the Forestry Agency, 1967-70; Mori 1983, 268). Regardless of which organization manages the national forest, committed and motivated personnel are vital.

The fourth and final problem concerns national forest legal matters. Until the 1998 reform, the 1951 National Forest Land Law (Law No. 246) governed the acquisition, maintenance, preservation, utilization, and disposition of national forest land. This law corresponds to a special case of the 1948 National Property Law (Law No. 73). The National Forest Special Accounting Law specified the management rules based on special accounting. The Special Measures Law of the National Forest Management Reform specified national forest financial reforms. The National Forest Administration Regulation, an instruction of the Ministry of Agriculture, Forestry, and Fisheries, was the most important legal provision relating to national forests. The forest planning system and management restrictions in non-national forest cases were specified by the Forest Law, but not those of the national forest. The Japan Federation of Bar Associations (1991) was critical of the fact that no basic law on national forest management existed.

For example, the Regional Management Plans and Forest Ecosystem Protection Areas, administered by the director general of the regional forestry office, lack any legal authority and are left to the arbitrary determination of the administrative organizations. Here the basic problem is that the Forestry Agency has two sides, the national forest management organization and a general forest policy planning agency. From the viewpoint of certification of free management activities under a self-supporting special accounting system, it appears more appropriate for a national forest manager to leave the determination of such an important management plan or protected forest (*hogorin*) to an administrative organization. The major role expected of national forests is changing from timber production to public benefit. Aid from the national general account has been increasing because of the national forest's financial difficulties. Under such conditions, management policy and plans relating to national forests must be determined and practised on a clear legal footing.

At the time of the 1998 national forest organizational reform, the laws relating to the national forest were also changed. The most important legal change was the amendment of the 1951 National Forest Land Law, which determines the acquisition, maintenance, preservation, utilization, and disposition of national forest land. In the 1998 reform, basic policy on national forest management was added to the contents of the law, and the title of the law was changed to National Forest Land Management Law. Article 3 lists the three goals of national forest land management: (1) realization of non-timber value, (2) timber supply, and (3) contribution to regional development.

The forest planning system with regard to the national forest was also changed in the 1998 reforms. The National Forest Basic Management Plan (planning period is 10 years), for the entire national forest, is based on the National Forest Land Management Law. Furthermore, harmonization with the National Forest Plan, based on Article 4 of the Forest Law (Law No. 249 of 1951), requires that both national forest and non-national forest be included. The 1991 Amendment to the Forest Law also introduced a Regional Forest Plan for the national forest (before 1991, Regional Forest Plans had been made only for non-national forest by the forestry departments of prefectural governments). In addition, a new plan called the National Forest Regional Management Plan (with a planning period of five years) was begun, based on the National Forest Land Management Law. This plan would determine the basic management policy for planning areas within the Regional Forest Plan.

A National Forest Operation Plan (with a planning period of five years) has been made based on Regional Forest Plans for the national forest and on the National Forest Regional Management Plan. This plan includes such forest practices as planting and cutting, and constitutes the smallest unit of forest planning in the national forest. It is not based on the National Forest Land Management Law but on an instruction of the Ministry of Agriculture, Forestry, and Fisheries.

The first National Forest Basic Management Plan, effective from 1 January 1999, was published on 25 December 1998, and the National Forest Regional Management Plan and National Forest Operation Plan were constructed by 31 March 1999. As a result of the 1998 reforms, the requirements of public notice and public inspection of national forest plans were added to the planning process, in order to make them properly available to the public. In the case of the National Forest Basic Management Plan, the draft plan was open to the public from 18 November to 18 December, and 44 responses (175 opinions, including similar opinions recorded) were sent to the Forestry Agency (Forestry Agency 1999a, 23-25). In the case of the National Forest Regional Management Plan and the National Forest Operation Plan, the regional forest offices of the Forestry Agency have to consider input from the governors of relevant prefectures and the mayors of relevant cities.

Since the 1998 reform, the national forest has been managed under a new accounting system, a system that includes general accounting. It is not clear at present, following the reform, just how the national forest will be managed in the future. It is notable, however, that the key phrase of the new national forest management is "Forest for People." The Forestry Agency (1998b, 29-32) has pointed out the following obligations: (1) disclosure of information on national forest management and the fulfillment of accountability procedures for national forest management, (2) promotion of public participation in national forest management, and (3) supply of non-timber value to the nation, and education of the public regarding national forest management and its basic policy directions in order to make national forest management accessible to the public.

The Forestry Agency has to reinforce the activities in accordance with the three main policy directions. Regional and district forestry offices of the Agency are also expected to modify their public services and forest practices in accordance with this new policy.

References

Forestry Agency (1946-96). *Kokuyu rinya jigyo tokeisho,* nos. 16-48. Forestry Agency, Tokyo (in Japanese).

– (1986). *Kessan jimu,* 26-33. Rinya-Kosaikai, Tokyo (in Japanese).

– (1990). *Shinrin no ryuiki kanri system,* 147-53. Nihon-Ringyo-Chosakai, Tokyo (in Japanese).

– (1994). *Ringyo hakusyo,* fiscal 1993 ed., 192. Nihon-Ringyo-Kyoukai, Tokyo (in Japanese).

– (1997a). *Ringyo hakusyo,* fiscal 1996 ed., 74-91, 141-44. Nihon-Ringyo-Kyokai, Tokyo (in Japanese).

– (1997b). Abstract of 1998 draft budget of Forestry Agency, *Rinya-jiho* 44(7): 2-17 (in Japanese).

– (1998a). "Heisei 8 nendo kokuyu rinya jigyo tokubetsu kaikei kessan no gaiyo," *Rinya-jiho* 45(7): 18-22 (in Japanese).

– (1998b). *Ringyo hakusho,* fiscal 1997 ed., 29-32, apps. 7-8. Nihon-Ringyo-Kyokai, Tokyo (in Japanese).

– (1999a). *Kokuyu rinya jigyo no bapponteki-kaikaku.* 439 pp. Nihon-Ringyo-Chosakai, Tokyo (in Japanese).

– (1999b). *Ringyo hakusyo,* fiscal 1998 ed., 117-33, 196-200. Nihon-Ringyo-Kyokai, Tokyo (in Japanese).

– (1999c). "Kokuyu rinya no keiei kanri ni kansuru kihon keikaku no jisshi jyokyo," *Rinya-jiho* 46(7): 19-27 (in Japanese).

– (2001). *Ringyo hakusyo,* fiscal 2000 ed., 212-14. Nihon-Ringyo-Kyokai, Tokyo (in Japanese).

Japan Federation of Bar Associations (1991). *Shinrin no asu wo kangaeru,* 230-52. Yuhikaku, Tokyo (in Japanese).

Maki, Shigehiro (1965). "Kokuyu rinya jigyo no tokubetsu kaikeihou rippoushi," *Ringyo Keiei Kenkyusyo Hokoku* 64(3): 102-107. Forestry Agency, Tokyo (in Japanese).

Management and Coordination Agency, Administrative Inspection Bureau (1990). *Kokuyu rinya jigyo no bapponteki kaikaku ni mukete,* 95-96. Printing Bureau, Ministry of Finance, Tokyo (in Japanese).

Mori, Iwao (1983). *Sugao no Kokuyurin.* 343 pp. Kosaido-shuppan, Tokyo (in Japanese).

Nomura, Nobuyuki (1956). *Ringyo kigyo keitairon.* 411 pp. Asakura-shoten, Tokyo (in Japanese).

Oka, Kazuo (1990). "Kokuyurin keiei." In Handa, R., ed., *Rinseigaku,* 2nd ed., 141-56. Buneido Shuppan, Tokyo (in Japanese).

Ouchi, Tsutomu (1989). "Kokuyurin mondai no shoten." In Nihon Rinsei Journalist no Kai, ed., *Watashitachi no mori kokuyurin wo kangaeru,* 63-83. Seibun-sha, Tokyo (in Japanese).

Shioya, Tsutomu (1978). *Rinseigaku,* 345-46. Chikyu-sha, Tokyo (in Japanese).

Tabuchi, Yuichi (1997). "Zaimu kara mita kokuyurin" ("National forest from a financial aspect"), *Ringyo Keizai (Forest Economy)* (587): 20-31 (in Japanese).

7
Forest Planning
Koji Matsushita

This chapter presents an overview of Japan's forest planning system and discusses its history, current program, and problems within its forest protection and planning systems. Japanese forestry has been guided by three major forest laws, often called the First (1897), Second (1907), and Third (1951) Forest Laws. There have also been minor amendments to these laws (e.g., the 1939 Amendment and the 1957 Amendment), which will also be discussed in this chapter.

The highest ranking plan among many plans related to land resources is the National Land Use Plan (hereafter called the National Plan) under the National Land Utilization Law (Law No. 92 of 1974). Article 2 of this law provides that the fundamental principles for Japanese land utilization planning are to promote: (1) preservation of healthy and culturally rich living environments, (2) balanced development of national land with preference for public welfare, and (3) conservation of natural environments with consideration for natural, social, economic, and cultural conditions. The National Plan consists of national, prefectural, and municipal plans. The current National Plan (the third plan) was approved by the Cabinet in February 1996 (the first plan was approved in May 1976 and the second plan in December 1985). The current plan's basic concepts include: (1) national land utilization emphasizing the forests' total management of watershed and soil conservation functions, (2) sustainable national land use in harmony with nature, and (3) national land utilization with beauty (National Land Agency 1996).

The Basic Land Use Plan (hereafter called the Basic Plan) is determined under the National Plan and involves city, agricultural, forest, natural parks, and natural environmental conservation areas. There are related laws for each area, including the Third Forest Law (Law No. 249 of 1951) for forest areas. The National Land Utilization Law is expected to play a harmonizing function as a national law under which related laws are enacted. As specified by the Basic Plan, the areas of each category at the end of FY 1998) are 26.7% city, 46.4% agricultural, 68.4% forest, 14.4% natural parks, and 0.3% natural environmental conservation (National Land Agency 1999). (The total exceeds 100% due to plural specification, where some areas are included in more than one category.) The Basic Plan specifies the area and clarifies the principles for coordination and administrative guidance.

Forests cover approximately two-thirds of Japan, including areas of plural specification under the Basic Plan. Forest plan areas required by the Forest Law must, naturally, adhere to the Forest Law. The history and current forest planning system are described in the next section. Japan's forest planning system is the most important aspect of its Forest Law. The forest protection system, described in the second section, is another important legal regulation comprising an important part of the Japanese Forest Law. Problems of the forest planning and protection systems are discussed in the last section.

Forest Planning System

Before the Second World War

The First Forest Law (1897) consisted of general provisions and forestry practice, protection forest, forest police, penalty, and miscellaneous regulations. Regulation of forestry practices received great attention. The law's most important objective was a continuous timber supply, harvest control, and land reclamation. Background for enacting the new law during the 1880s and 1890s follows Handa and Ariki (1996).

In the 1896 bill that was not adopted, regulations of forest practice were applied to all non-national forest land. The National Diet, however, opposed such broad regulatory application, and with the First Forest Law limited the regulations to public forests and to shrine and temple forests. The First Forest Law also gave prefectural governors the power to specify forest management methods when lacking sustainability for these forests. In the 1907 Second Forest Law, the governor could authorize management principles or a plan for public and shrine and temple forests, when required. Both the 1897 and 1907 Forest Laws targeted public and shrine and temple forests, including regulations to prevent the denuding of forestland by the former and to promote forest management to enhance timber production by the latter.

There were many denuded forestlands, as well as increasing social concern over water conservation and erosion control, when the national government began managing national forest land. To promote forest management, the Second Forest Law added two new chapters concerning utilization and expropriation of land, and forest owners' associations. Government authority was widely strengthened in the 1939 Amendment.

In this amendment, owners of all non-national forest land other than public and shrine and temple forests began creating management plans that required authorization from the governor. Owners holding less than 50 ha of forest joined in collective forest management plans prepared by forest owners associations. The governmental supervision system of forest management planning was considered completed with the 1939 Amendment, since the obligation to prepare forest management plans was finally extended to all non-national forest land. Despite being completed as a legal system, fundamental forest management functions of the planning system functioned poorly because of overcutting during and immediately following the war.

After the Second World War
With passage of the Third Forest Law (1951), a government supervision system of forest management replaced the forest planning system. The Forestry Division, Natural Resource Section of the General Headquarters of the Supreme Commander for the Allied Powers was very interested in the government supervision system and ordered it continued (Shinrin Keikaku Kenkyukai 1992, 4). Since the administrative system requiring forest owners' associations to make forest management plans (1939 Amendment) violated the anti-monopoly law, a basic occupation policy, the government started a new forest planning system that made forest owners obey the forest management regulations (Shinrin Keikaku Kenkyukai 1992, 4).

The forest planning system of the Third Forest Law consisted of a nationwide Basic Forest Plan, Forest District Management Plans, and Forest District Working Plans for non-national forests (including both private and public forests), and Managerial Plans for national forests.

The Basic Forest Plan, designated by the Minister of Agriculture and Forestry, was a five-year plan for the forest area in each Basic Plan district. There were 376 planning districts designated by the Forestry Agency in 1951. The Forest District Management Plan was a five-year plan for non-national forest land in each forest district (2,096 districts in 1951) prepared by the governor of each prefecture within the planning area. The Forest District Working Plan, also prepared by the governor, was an annual working plan for non-national forest land in each forest district. For national forest land, the Managerial Plan was a five-year plan prepared by the director general of the regional forest office for each forest management unit (546 in 1951). For non-national forest in areas other than unrestricted forest, all harvesting done beyond the normal cutting age required prefectural government permission. Moreover, forest owners were obliged to replant after cutting, and the planting locations were specified in the forest planning system.

With the 1957 Amendment, a harvest-reporting system replaced the permission system so that on non-national forest land, harvesting of broad-leaved trees in unrestricted forests no longer required permission. Forest owners were required to report their cutting plans during a specified period before starting to cut. A forest management planning system for public forests was also included.

In the 1962 Amendment, the annual Forest District Working Plan was abolished. All cutting activity among unrestricted trees was controlled under the harvest-reporting system. The government must determine the Long-Term Perspective on Demand and Supply for Important Forest Products (hereafter called the Long-Term Perspective) and the Basic Plan on Forest Resources (hereafter called the Basic Forest Plan) under the Basic Forestry Law (Law No. 161 of 1964). According to these two plans determined by the government, the Minister of Agriculture and Forestry must develop the National Forest Plan (similar to the former Basic Forest Plan system), but with the planning period extended from five to 10 years. Regarding the non-national forest planning system, the former system (Forest District Management Plan) was changed to a new regional forest plan and the planning area under the new system was expanded to include the former system's basic plan district area. The compulsory characteristics of the postwar forest planning system were reduced considerably by the 1957 and 1962 Amendments.

The 1968 Amendment introduced a new forest planning system, the Forest Operation Plan, that required individual forest owners to develop management plans for their forests with approval from the prefectural governor. The Territorial Joint Forest Operation Plan was introduced in the 1974 Amendment. Japanese privately owned forests are generally small and scattered in location. Conducting effective forest management requires a minimum forest area. Thus, when each private forest ownership's total forest area exceeded 30 ha and the forest satisfied certain conditions, forest owners in that area could develop a joint Forest Operation Plan. The 1974 Amendment introduced the River-Based National Forest Plan to the National Forest Plan by dividing Japan into 29 areas based on major rivers. The Forest Land Development Permission System was introduced due to lack of an effective policy controlling harvesting in the unrestricted forests of non-national forest land. The 1983 Amendment introduced a new forest plan, the Forest Improvement Plan, that focused on tending and thinning. It included a new concept that the plan was to be developed by municipal offices. The forest planning system changed again when the 1991 Amendment introduced to forest policy the basic concept of "watershed" forest management.

The Current System
The characteristics of the current forest planning system (Figure 7.1), which was introduced in the 1991 Amendment, are summarized here:

Figure 7.1

Forest planning system introduced in the 1991 Amendment to the Forest Law.

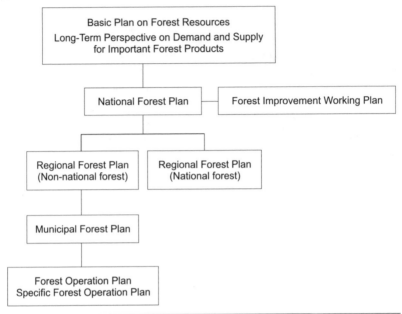

- The National Forest Plan is to be revised every five years, within a planning period of 15 years. The planning area based on the system of main rivers changed from a base of 29 to 44 areas, with each planning unit occupying less area. Another important change, indicating its greater importance, is that the National Forest Plan is now approved by the Cabinet.
- The Forest Improvement Working Plan was created and added to the forest planning system as the financial basis of the National Forest Plan. This new five-year plan strives to improve afforestation and forest road construction nationwide. Although such a long-term improvement plan is basic for the success of the forest planning system, there had been no system addressing the financial base of Japan's forest planning system.
- The Forest Improvement Agreement was introduced. It is to be developed between governors or mayors of the upper and lower reaches of the same river, aiming through discussion and mediation to foster cooperation among autonomous local entities in developing forest improvement works.
- The planning area of the Regional Forest Plan was reconstructed around a total of 158 new planning units. Previously, separate Regional Forest Plans

were developed by the prefectural government for non-national forest land and by the Regional Forestry Office for national forests. The 1991 Amendment required the prefectural government and local forestry offices to prepare the 10-year Regional Forest Plans cooperatively, revising them every five years.

- The Forest Land Development Permission System was partially changed, involving the conditions that require permission for forestland development.
- The Municipal Forest Plan was changed to the Municipal Forest Improvement Plan, a 10-year plan to be revised every five years. The main planning issues under the previous system were tending and thinning activities in each municipality. Joint work on forest management and the promotion of the mechanized forest operation are added under the new improvement plan. With the aim of disaster prevention, the new system permits municipal offices to do thinning and tending of forests for owners living outside the municipal territory.
- A new type of Forest Operation Plan, the Specific Forest Operation Plan, was added, to improve the non-timber values on non-national forest lands using the new no clear-cut (i.e., multi-storied operation) forest management system.

Achievement of a Forest Planning System

Since 1964, under the Third Forestry Law, the Long-Term Perspective and Basic Forest Plan make up the top of the forest planning system (Figure 7.1). The Long-Term Perspective is a report including future trends of timber demand and supply clarified using econometric and statistical analysis. The National Forest Plan must adhere to the Long-Term Perspective and Basic Forest Plan. Data quantifying three major objectives, evaluated in seven National Forest Plans, are shown in Table 7.1. Although National Forest Plans include much more data, we focus here on three important indexes: cutting volume, artificial planting area, and length of forest road construction. Table 7.1 presents annual values calculated for the total planning periods. Since the intervals of the National Forest Plan renewal were not equal to the planning period, "ratio of achievement" values have been averaged between the plan's starting year and the previous year of the next plan's starting year.

Both the "planning values" and "actual results" of cutting volume and artificial planting area trend downward. The actual results of cutting volumes are approximately 60-80% of the planning values for all plans. Regarding planting area, the actual results were almost equal to the planning values in the first and second National Forest Plans. After the fourth plan, however, the percentages of actual results versus the planning values had fallen to approximately 50-60%, similar to cutting volumes. Regarding length of forest road construction, the planning values peaked in the fourth plan

Table 7.1

Evaluation of National Forest Plans in Japan with regard to three
indexes: cutting volume, area of artificial forests, and length of forest
road construction.

Item	Plan no. of National Forest Plan	Planning period (fiscal year)[1] Start	End	Volume of cutting[2]	Area of artificial forests[3]	Total length of forest roads[4]
Planning	1	1963	1972	82	418	5.0
values	2	1968	1982	82	357	5.8
	3	1973	1987	65	325	8.6
	4	1978	1992	65	275	9.2
	5	1983	1997	64	218	6.0
	6	1988	2002	49	117	4.5
	7	1991[5]	2006	46	93	4.4
	8	1997	2011	40	67	3.8
Actual	1	1963	1967	74	377	3.0
results	2	1968	1972	65	338	2.7
	3	1973	1977	46	231	1.8
	4	1978	1982	42	167	2.7
	5	1983	1987	40	108	2.0
	6	1988	1991	38	69	1.9
	7	1992	1996	31	51	1.3
	8	1997	1999	26	38	1.7
Ratio of	1	1963	1967	90	90	60
achievement	2	1968	1972	79	95	46
(%)	3	1973	1977	71	71	21
	4	1978	1982	65	61	29
	5	1983	1987	63	50	34
	6	1988	1991	78	59	42
	7	1992	1996	67	55	30
	8	1997	1999	65	57	45

1 In the case of planning values, "period" refers to planning periods. In the case of actual
   results and ratio of achievement, "period" refers to periods for which actual results are
   calculated.
2 Annual average volume in million cubic metres.
3 Annual average area in thousands of hectares.
4 Annual average length in thousands of kilometres.
5 Started on 9 August only in the seventh plan.
Sources: Shinrin Keikaku Seido Kenkyukai (1963), *Practical Business of Forest Planning (Shinrin
keikaku no jitsumu);* Shinrin Kihon Keikaku Kenkyukai (1997), *Long-Term Vision on Forests and
Forestry for 21st Century (21 seiki wo tenboushita shinrin ringyo no tyouki vision);* Forestry Agency
(1970), *Practical Business of Forest Planning (Shinrin keikaku no jitsumu);* Forestry Agency (1974,
1979, 1995), *Compendium of Laws on Forests (Rinya sho-roppo);* Forestry Agency (1984, 1988),
Forest Planning Handbook (Shinrin keikaku gyomu hikki); Forestry Agency (1967-2001), *Statistical
Handbook of Forestry (Ringyo tokei yoran).*

and decreased thereafter. The percentage of actual results versus planning values was 60% in the first plan but less than 50% thereafter.

Table 7.1 reflects large differences between the actual and planned values in all three indexes. One reason is that the forest planning system has no financial base. Thus, the Forest Improvement Working Plan, a five-year plan to improve plantation and forest road construction, was introduced in the 1991 Amendment. This plan quantifies the amount of plantation and forest road construction necessary to accomplish the objectives of the National Forest Plan. Since the Forest Improvement Working Plan is a long-term government public works plan that must be linked to the national budget, the working plan is a Cabinet matter, just like other plans for public works related to the Ministry of Agriculture, Forestry, and Fisheries. These include agricultural land improvement projects, erosion control projects, water regulation projects, and fisheries improvement projects. In April 1992, the first Forest Improvement Working Plan, which covered fiscal years (FY) 1992 to 1996, was formulated, and the total spending for these works improvements was ¥3,900 billion (US$32.5 billion; US$1 = ¥120), an annual budget of ¥780 billion.

After the National Forest Plan was revised in 1996, the second Forest Improvement Working Plan, which covers FY 1997 to FY 2003, was implemented in December 1997. The total spending for these works improvements was ¥5,380 billion (US$44.8 billion; US$1 = ¥120), an annual budget of ¥769 billion. The annual budget was slightly smaller than in the first plan, because of the poorer condition of the national finances.

The 1996 Revision of the Basic Plan on Forest Resources
The Basic Forest Plan, Long-Term Perspective, and National Forest Plan (planning period: 1 April 1997 to 31 March 2012) were revised in 1996. The revised Basic Forest Plan included three important changes (Tsumoto 1997). First, it introduced sustainable forest management. In addition to conventional policy measures based on the quantitative views (e.g., forest area and volume), the new plan included quality improvement (e.g., ecosystem health plus soil and water conservation). The preparation of forest resource data is also discussed, in accordance with the current development of an international agreement on the definition of sustainable forest management.

Second, a change in the classification method for forest area was specified. Since around 1955, the basic classification criteria were "artificial" (*jinkorin*) and "natural" forests. Before this, forests were classified as coniferous forests, broad-leaved forests, and mixed forests. In the new classification method, forest area is first divided into two categories, "plantation" (*ikuseirin*) and "natural" forests. Plantation forests are managed artificially, including supplementary planting or soil scarification in the natural forest, and are then subdivided into single- and multi-storied forests. The new system attaches importance to the degree of artificial intervention in the forest and the forest stand structure, including species.

Third is the deletion from the new plan of previously important policy objectives in expansive afforestation (conversion of coniferous forestland previously used for fuelwood plantations to plantations whose timber is used mainly for construction materials). In the new plan, plantations on treeless land and forestland damaged by disease and insects are regarded as important.

As these new ideas have just been introduced to the Basic Forest Plan, the methods of realizing such important new objectives at the regional or individual forest plan level are still developing.

The 1998 Amendment to the Forest Law

The 1998 Amendment to the Forest Law includes four major changes that cover thinning, Specific Forest Operation Plans, municipal policy on forests and forestry, and public involvement in forest management. Revisions in these areas were first introduced as basic forest policy in the 1991 Amendment, although there were improvements in the 1998 Amendment. The changes are summarized in the following paragraphs (Hashimoto 1998).

To promote thinning, two points were changed. First, the approval of thinning was added to the Forest Operation Plan. Previously, approval under the Forest Operation Plan focused mainly on planning forest practices relating to the final harvest and sustained yield, and there were no clear approval criteria for thinning, although it was recognized that thinning was necessary. In a recent survey of non-national forest plantations, however, it became clear that the proportion of forest areas where thinning was practised was only approximately 50% of the total forest area where thinning was considered necessary (Forestry Agency 1998). The forest area covered by the Forest Operation Plan, which was introduced in the 1968 Amendment, was 13.5% in FY 1970, 54.6% in FY 1980, 71.8% in FY 1990, and 74.2% in FY 1996. To promote the proper thinning of artificial forests, guidelines on thinning were added to the approval criteria of the Forest Operation Plan covering non-national forests. The second change concerns thinning in protection forests (*hoanrin*). Previously, all cutting activities in protection forests, including thinning, were regulated by the permission system. In the 1998 Amendment, a harvest-reporting system replaced the permission system for thinning, mainly because it simplified the procedure and promoted thinning.

The Specific Forest Operation Plan was first added to the forest planning system in the 1991 Amendment. Two points were changed in the 1998 Amendment. First, only individual owners could make Specific Forest Operation Plans according to the 1991 Amendment, but the 1998 Amendment includes joint planning by forest owners. In the case of Forest Operation Plans, the percentage covered by Territorial Joint Forest Operation Plans increased to 87.1% in FY 1996, that is, most of the Forest Operation Plans were made by more than one forest owner. In the 1998 Amendment, the territorial joint planning method was extended to include Specific Forest Operation Plans. The second change related to the type of forest targeted by Specific Forest Operation Plans. The forests targeted were limited to artificial forests in the 1991 Amendment, and natural forests were included in the 1998 Amendment. The aim is to promote the management of broad-leaved forests, with the objective of growing more scenic forests in rural areas.

The scope of Municipal Forest Plans, which are formulated by municipal offices, was extended in the 1991 Amendment. The role of municipal departments of forestry in making policy, including Municipal Forest Plans, was further strengthened in the 1998 Amendment. Three changes were made.

First, only 2,020 of the 3,058 existing municipalities were required to produce a Municipal Forest Plan under the 1991 Amendment. In the 1998 Amendment, all municipalities that have private forests within their jurisdiction must make a Municipal Forest Plan. Second, the Municipal Forest Plan was expanded to cover all forest practices related to non-national forests, from planting to cutting, within their jurisdictions. This basic policy was already introduced in the 1991 Amendment, and the policy became clearer in the 1998 Amendment. With the expanded scope of Municipal Forest Plans, the focus of the Regional Forest Plan (formulated by the prefectural government) was limited to clarifying basic forest management policy for forest planning areas based on "watersheds." Third, in the 1998 Amendment, municipalities were authorized to control non-national forest management. The main responsibilities transferred to municipalities were to make recommendations for forest practices, to accept harvest reports, except for protection forests, and to approve Forest Operation Plans and Specific Forest Operation Plans, which were formerly authorized by prefectural governments.

Recently, because of the worsening economic picture of the domestic forestry industry and the increasing tendency to add non-timber values to forest resources, the importance of including the general public in forest resource management and the forest planning process has increased. The 1991 Amendment introduced a new system of managing non-national forests involving cooperation between municipalities located on the upper reaches of a river (forest area) and those on the lower reaches (non-forest area). To promote further public participation in forest management, the following two points were changed in the 1998 Amendment. First, initial public hearings were added to the planning procedure for Regional Forest Plans and Municipal Forest Plans. Previously, the public hearing procedure began after the publication of these plans. Second, the cooperative system of forest management between municipalities, which was introduced in the 1991 Amendment, was improved. Formerly, it focused mainly on public forests and the measures used to achieve cooperation were the introduction of a profit-sharing system and the establishment of a specific organization to deal with profit-sharing forests. In the 1998 Amendment, the cooperative system was expanded to include private forests, and a funding system to assist private forest management was added.

The 1998 Amendment includes the four major changes outlined above. With the realization of the 1996 revision of the Basic Plan for Forest Resources and the National Forest Plan, the extension of these new ideas and actual planning based on new ideas are future problems for the forestry sections of the national, prefectural, and municipal governments.

Forest Protection System

Overview

Protection forests are called *hoanrin* in Japanese. The *hoanrin* system, a major component of the Japanese Forest Law and the forest planning system, covers both national and non-national forests. Beginning with a brief history, this section discusses the forest protection system.

At the beginning of the Meiji period (1868-1912), there were periods when forest policy was ineffective and the area of denuded forests increased. There was considerable flood damage in the 1890s. Thus, the major issue addressed in the First Forest Law (1897) was development of the forest protection system, including 12 kinds of protection forests. To prevent natural agricultural calamities and protect the environment in the Edo period (1603-1867), there was already a protection system similar to this new forest protection system. The First Forest Law provided a legal base and unified name for such traditional forest protection. The Second Forest Law further developed the forest protection system as a policy system.

In the Third Forest Law (1951), the base of current law, the kinds of protection forests increased from 12 to 17 (see Table 7.2). Japanese Forest Law did not define "protection forest" directly, but listed the 17 types of protection forests based on their expected roles. Headwater-conservation forests, soil-loss prevention forests, and crumbling-soil prevention forests are very important and are thus collectively called Watershed-Conservation Protection Forests. Except for these three and recreation forests, the geographical distribution of each protection forest area is limited. Snow prevention forests have an extremely small area and a very limited specified location. Fog prevention forests are located mainly in southeast coastal Hokkaido. Fish-breeding forests are concerned with the contribution to fish habitat and breeding by the forest shadow reflected on the water or the forest's water pollution prevention function. Navigation-target forests, concerned with maintaining safe navigation, are used as navigation targets and/or landmarks by fishing boats. Fish-breeding forests and navigation-target forests have been determined for the benefit of the fishery industry. Recreational and scenic-beauty protection forests are specified to promote social welfare.

Table 7.2

Types and areas of protection forest (*hoanrin*) in Japan, 1951-99.

Type of protection forest[1]	Area (1,000 ha)[2]						% in 1999
	1951[3]	1960[3]	1970[3]	1980[3]	1990[3]	1999[3]	
Headwater-conservation forests	1,273	2,191	5,130	5,419	6,028	6,387	67.4
Soil-loss prevention forests	886	1,076	1,438	1,621	1,936	2,103	22.2
Crumbling-soil prevention forests	21	37	41	45	46	51	0.5
Shifting-sand control forests	15	14	14	16	16	16	0.2
Windbreak forests	69	49	51	54	55	56	0.6
Flood control forests	3	3	1	1	1	1	0.0
Tide water control forests	7	9	8	12	13	13	0.1
Drought disaster control forests	3	11	20	32	40	73	0.8
Snow prevention forests	0	0	0	0	0	0	0.0
Fog prevention forests	31	51	55	53	51	59	0.6
Avalanche prevention forests	12	14	14	18	19	19	0.2
Stone fall prevention forests	1	1	1	2	2	2	0.0
Firebreak prevention forests	0	0	0	0	0	0	0.0
Fish-breeding forests	46	36	29	28	28	29	0.3
Navigation-target forests	1	1	0	1	1	1	0.0
Recreation forests	0	0	1	158	554	635	6.7
Scenic-beauty forests	37	29	23	29	28	27	0.3
Total[4]	2,406	3,522	6,829	7,489	8,818	9,473	100.0
Total (2)[4]				7,317	8,297	8,867	

1 The types of protection forest are listed in order of rank as prescribed in Article 25 of the Forest Law.
2 Before 1970, if a forest is designated under two or more categories, it is listed in the highest ranking category among them. After 1980, if a forest is designated under two or more categories, it is listed in each category.
3 End of fiscal year.
4 Area figures have been rounded, so "Total" differs slightly from the sum of individual protected forest areas. After 1980, a forest designated under two or more categories is counted in each category, so the total area exceeds the real total area of protection forests. The real total areas are shown in the "Total(2)" row.
Sources: Forestry Agency (1953, 1962, 1972, 1982, 1992), *Statistical Yearbook of Forestry (Ringyou toukei youran);* Forestry Agency (2001), *Ringyo Hakusyo,* fiscal 2000 ed.

Enacted in 1954, the Temporary Measures Law for Protection Forest Consolidation was intended to be valid for only 10 years. Nonetheless, it continues under a 10-year planning period and renewal cycle. The Consolidation Plan for Protection Forest (hereafter called the Consolidation Plan) was the basis for this Temporary Measures Law. The main aims and characteristics of each 10-year plan were as follows (Fukao 1996):

In the First Consolidation Plan (1954-63), the main objective was to promote the improvement of protection forests to prevent natural calamities. Since the comprehensive promotion of watershed conservation was the main objective, Japan was divided into 216 planning areas on a watershed basis. The government decided to purchase important forestlands related to land conservation, generally inaccessible water-source forests.

In accordance with the increasing water demand due to the rapidly expanding Japanese economy, the Second Consolidation Plan's (1964-73) main objective was the expansion of Watershed-Conservation Protection Forests, especially headwater-conservation forests.

The Third Consolidation Plan's (1974-83) main objectives were the expansion of recreational forests and countermeasures against deterioration of the natural environment from urbanization and increasing demand for recreational use of forest areas. Through these three periods of consolidation, the quantitative development of the forest protection system was almost completed.

In the Fourth Consolidation Plan (1984-93), forests for local disaster prevention (crumbling-soil prevention forests) were increased to protect residents in mountainous regions and were linked with erosion control projects. In 1984, the Specified Protection Forest System was introduced, allowing the Minister of Agriculture, Forestry, and Fisheries to specify the forests (mainly non-national forests) where expected roles are decreasing due to the lack of necessary forest practices as Specified Protection Forest. For the purpose of recovering these functions under the forest planning system and the plan made by the prefectural government for the Specified Protection Forest (e.g., plantation method), the prefectural governor can specify forest practices.

The Fifth, and current, Consolidation Plan (1994-2003) gives priority to local disaster prevention, demands for high-quality water, national demands for environmental resources, and the necessity for disaster prevention in inadequately managed forest areas (Forestry Agency 1994).

As a result of these plans, the protection forest area at the end of FY 1999 had increased by 3.7 times over that in 1951, reaching 8,867,000 ha (35.3% of the total forest area at the end of FY 1994). The areas of headwater-conservation forests and soil-loss prevention forests comprise 67.4% and 22.2%, respectively, of all protection forest area, and the combined total area of both types of protection forests is approximately 90%. Of all protection forests, the area of headwater-conservation forests has increased the most since the war, followed by those of soil-loss prevention forests and recreation forests.

Specification and Removal of Protection Forests

Specification of protection forest areas is based on application by interested persons and/or determination by the government on the basis of need. The latter has been more common in postwar protection forest specification, based on Consolidation Plans. When a forest area is expected to perform multiple functions, the same unit may be assigned to more than one type of Protection Forest (plural specification). The Minister of Agriculture, Forestry, and Fisheries has the authority to specify and remove protection forest designation for headwater-conservation forests, soil-loss prevention forests, and crumbling-soil prevention forests. Prefectural governors may specify other kinds of protection forests. Based on the Third Forest Law, the coastal conservation district (established by Coastal Law, Law No. 101 of 1956) and the strict natural environment protection area (established by the Natural Environment Protection Law, Law No. 85 of 1972) can not be specified as protection forests. There are three important issues regarding protection forest specification and removal.

The first issue relates to the types of protection forests. Although under current law the types have remained unchanged since 1951, administrative inspection (1964) has revealed that the importance of several types has declined considerably as their roles have changed. For example, while snow prevention forests are classified by law, only an extremely small area of forest is actually so designated. Practically, however, railways (especially in northern Japan) use forests adjacent to their lines in a "snow protection" role. Fish-breeding forests provide another example. The economic roles of coastal fisheries seem to be in general decline. Recently, however, several fishery-related organizations prepared funds for upstream forest management. Thus, despite changing social conditions, the legal regulation of fish-breeding forests remains unchanged. The role of navigation-target forests also seems to be decreasing due to the technical development of ships. The importance of recreation forests has rapidly increased due to urban expansion and increasing demand for forest-based tourism. Thus, the role of recreational forests must be reconsidered and further developed. Generally, in the current forest protection system, roles related to land conservation and disaster prevention are seen as highly important.

The second issue arises as a result of the rapid expansion of specific areas after the war. Most protection forests were specified by the government. Areas specified due to requests by forest owners or other interested people, such as firebreak prevention forests, were small. The specified area has increased due to both administrative and forest-owner reasons. In general, the national and prefectural governments were aiming to increase the protection forest area, while forest owners who agreed with the protection forest specification failed to adequately estimate the importance of their forests' non-timber values. Their concern was in maximizing subsidy payments and tax cuts, the most important of which involved exemption from municipal property tax and abatements of inheritance and gift taxes. Area expansion was impossible without these adjustments to the taxation system. Generally, citizens who do not own their forests are not concerned with the protection forest specification nor do they understand the Protection Forest System. Accordingly, the post-war increase in protection forest area did not come about because forest owners and citizens had become more familiar with the varieties of forest resources. When the area specified as a protection forest becomes the target of a development plan (e.g., golf course, resort facilities, and housing construction), most forest owners want to remove the specification and sell the forestland. This attitude indicates that owners of protection forests do not recognize the various non-timber values of forest resources.

The third issue is related to the authority to remove the protection forest specification. When local public entities or recreational companies create development plans that include protection forests, this becomes a political issue. Generally, citizens opposing development tend to oppose the removal of the protection forest designation. In the case of Watershed-Conservation Protection Forests, which comprise a large area, the authority to remove the specification belongs to the national government, not the prefectural government. Thus, assigning authority to remove specification to the prefectural governor has been frequently discussed, and the discussions include arguments related to the decentralization of power.

Limitation of Forest Practices and Loss Compensation

Limitations are imposed on forest practices in protection forests, based on the type of protection forest involved. In the 1967 Amendment, cutting methods and quantitative limitations on cutting were included among specific forest practices, as were post-harvest planting methods, periods, and acceptable species. The protection forest type determines the cutting method permitted. At the end of FY 1996, 3.1% of protection forest area was specified as felling-prohibition forest while 20.5% of protection forest area permitted selective cutting and 76.4% permitted clear-cutting (Nihon Chisan Chisui Kyokai 1997, 987). When clear-cutting is permitted, there is a limitation on the size of the cutting area. For example, the limit is 20 ha for headwater-conservation forests and 10 ha for soil-loss prevention, shifting-sand control, and drought prevention forests. Limited to cases where trees exceed the standard rotation age, protection forest may be cut with permission.

Because forest practices are limited, there is a program to compensate owners for economic loss on privately owned protection forests. The compensation program is effective for the felling-prohibited forests and forests where the single-tree selection system is permitted. In felling-prohibited forests, 5% of the estimated value of standing trees over standard rotation age is paid as compensation every year in a program based on the concept of interest subsidy on frozen assets. Standing trees are evaluated every 10 years. Forest owners who accept the protection forest specification receive preferential considerations such as favourable loans, higher subsidies, and various tax cuts; these are the main reasons for accepting protection forest status.

One problem involved in compensation is obtaining funds for payment. Generally in Japan the budget for forest policy other than construction (such as forest road construction and dam construction) is insufficient to provide adequate compensation. When the site is a very important forest, it must be purchased by the national or prefectural government. Generally the procurement budget is quite small, making it difficult to increase the area specified as felling-prohibited forests. The fact that clear-cutting is permitted on 76.2% of protection forest areas demonstrates the existence of the funding issue and the tendency to specify protection forests where compensating for the owner's loss is unnecessary.

Another problem regarding loss compensation is the lack of a theoretically sound, understandable method of calculation. Certainly, specifying protection forests causes economic loss, but in general forests also include non-timber values. Generally these non-timber values are as important as continuous timber production, but it is difficult to quantify the economic loss as standards change over time.

Problems of the Forest Planning and Protection Systems

The Differences in Planning Values and Practices
As discussed earlier, there are large differences between several important predicted and actual planning values. There are three reasons for this serious problem.

The first is that the method for estimating future timber cutting areas and volumes is based on Gentan Probability Theory, a sort of Markov Chain Model where future cutting probabilities are calculated from past cutting activities. In Japan, however, the cutting volume has been decreasing since 1967 despite the increasing trend in total growing stock. During this time, because of the nature of forest growing stock, the Gentan Probability Theory overestimated cutting volumes. Thus, the quantity of forest resources available to cut has increased, whereas, in accordance with past activities, the cutting probability was assumed to be constant. In the calculation procedure, which considered only supply-side conditions, managerial conditions surrounding both forests and the national economy (such as timber price, labour cost, and forest road construction) were not considered at all. Blandon (1996) noted the statistical problem of the Japanese long-run timber supply model calculation process and discussed the method of applying Gentan Probability Theory actual estimations. Overestimated values for future cutting activities caused overestimated planning values for future planting areas and subsequent forest practices such as thinning. Overestimating the cutting volume was also related to overestimating the demand for forest road construction.

The second reason relates strictly to the fact that the forest planning system has been so highly important in forestry policy activity by the Forestry Agency. Thus, when it is decided as a basic forest policy framework that future forest practices (for example, harvesting, thinning, and road construction) will decrease, this prediction influences all Forestry Agency activities and the national budget distribution. Japanese administrative sectors in general, including the Forestry Agency, tend to overestimate in hopes of strengthening the organization by using inflated predictions to obtain higher budgets from the Ministry of Finance.

The third reason is the objective of the forest planning system. Governmental supervision of forest management before and during the Second World War carried strong compulsory power on non-national forest lands, while the planning system introduced by the current postwar Forest Law has important and strong policy measures (for example, the cutting permission system). The current planning system has almost no compulsory power over forest owners, however. Accordingly, several planning values on future forestry activities have limited importance for both administrative sectors and forest owners. In this sense, planning values are not predictions, so some see no real problem caused by differences between estimated and actual activities. There are problems, however, in the planning system itself because the planning system values have always been overestimated. Regarding cutting volume, for example, it is also problematic that the administrative offices appear to make almost no effort to achieve the objective (Akai 1984, 205).

The forest planning system has several levels (Figure 7.1). A more serious problem, including overestimated planning values for future forestry activities, occurs at the Regional Forest Plan level, which is based on the National Forest Plan. A balance is required between the planning values in the National Forest Plan and the total planning values in the Regional Forest Plan. Thus, great differences between estimated and actual situations involve almost all regional forest planning areas. As forestry administrators and forest owners better understand the regional forest planning area, the differences become clear and the planners must make a greater effort to distribute the overestimated future forestry activities among the planning areas.

Also an issue, such planning conditions have become the normal situation. This causes several other problems for everyone involved in forestry, such as decreasing the planner's effort to improve forest plans, decreased trust in the Regional Forest Plan by forest owners, and a decrease in the utilization values of the Regional Forest Plan by the non-forestry administrative planners. Finally, ineffective and perfunctory plans continue to be produced, preventing effective Regional Forest Plans from being formulated.

Thus, most forest owners do not understand the Regional Forest Plan governing their own forest management plan. Such understanding is unnecessary, however, because the Regional Forest Plan does not always reflect actual regional situations. While the forestry technicians who made the plan in accordance with the National Forest Plan understand the situation well, they do not positively promote the planning goals to regional forest owners and citizens.

Just after the Second World War, when unplanned cutting activities continued and there was much denuded forestland, the forest planning system functioned importantly in recovering forest resources, maintaining self-sufficiency with regard to timber supply and demand, and providing sufficient domestic forest resources. Recently, timber self-sufficiency planning has been unnecessary. When compared with the level required for sustained yield, there are no possibilities of overcutting. Basically, the conventional forest plan has been based on stringency in the future supply and demand situation of Japanese timber markets (Akai 1984, 207). Thus, the various objectives of artificial forests have been stated in the National and Regional Forest Plan. This assumption is quite different from both current market conditions and predicted future situations. Thus, it must be discussed whether or not it is appropriate for the current planning system to have national goals first set by the national government and then distributed to the regional planning areas, where the Regional Forest Plans are provided by prefectural governments.

Such a system is troublesome in practice. Overestimating the national goal risks promoting an impractical and excessive working plan at the Regional Forest Plan level. Recently for example, an increase in multi-storied forests was indicated in the previous National Forest Plan (planning period: 1991-2007). The planning values of the multi-storied forest area in the Regional Forest Plans were estimated by distributing the national goal of planning values (Table 7.1). Each regional forest planner must search for places to introduce new forest practices. At that time, if private forest owners are not interested in creating multi-storied forests, public or private forests of the representative regional forest owners are forced to implement the practices and show genuine results. Recently, the internal rate of return for artificial forest management has steadily declined. Although the national goal of developing areas of artificial forests disappeard in the 1996 National Forest Plan, plantation subsidies are still distributed to prefectural governments and private forest owners according to the forest plan. Consequently, the area of artificial forests will increase. The national forestry budget is influenced by an impractical forest plan and ineffectively distributed to local governments without concern for the real situation in regional forestry and forest resources. Further, such a budget distribution system cannot always cope with the improvement of forest resources and the development of forestry and related industries.

While one prefecture consists of several regional forest planning areas, one forest planning area includes many municipalities. Before 1991, the relationship between the forest planning system and the forest policy of municipal offices was unclear. The 1991 Amendment introduced the Municipal Forest Improvement Plan. Since municipal offices are closest to the residents and forest owners, they must be involved in the regional forest policy. Most municipal offices, however, lack any expert foresters or forestry sections.

Control of Forestland Development
The main policy measures controlling forestland development are the Forest Protection System and Forest Land Development Permission System. As previously indicated, under the current Forest Law all cutting activities in protection forests require permission. The Forest Land Development Permission System was introduced in the 1974 Amendment to control forestland development in the forest areas not specified as protection forest. Under the Forest Law, development of forestland such that the developmental work will influence the surrounding planning area in the Regional Forest Plan requires permission from the governor. Development is defined as all work that changes the form of the forestland, including gathering stones and stumps or beginning cultivation. The minimum size of forestland development requiring permission is 1 ha and is determined by government ordinance after considering the natural forestland conditions and the development characteristics.

In 1987, the Resort Area Development Law (No. 71) was enacted to promote comfortable living and the development of local economies by general improvement of recreation, sports, cultural, and meeting activities in areas prepared for tourism with lodging accommodations. Since recreational use of forest resources (for example, for sports and tourism) was not clearly defined in the forest planning system under the Forest Law, a new law, the Special Measures Law on Promotion of Recreational Function of Forest (No.

70) was enacted in 1989. Its objectives, more comprehensive than those of the Forest Law, are aimed at economic development of forested areas and improvement of the national welfare.

Although forest resources generally serve various functions, the Special Measures Law deals with recreation only. Under the law, following the revision of the National and Regional Forest Plans, the Minister of Agriculture, Forestry, and Fisheries develops the basic national policy promoting recreational use of forest resources. Under the law, control of forestland and the limitation of cutting activities are excluded from the Forest Law's permission system (namely, Forest Protection and Forest Land Development Permission Systems). From a legal perspective, these recreational development activities are considered exceptions to the Forest Law because they are already authorized with revisions of the National and Regional Forest Plans.

As Japan experienced an extraordinary economic boom in the latter half of the 1980s, there was strong pressure for forestland development. The area that was developed under the Forest Land Development Permission System was 20,178 ha in FY 1985; 17,899 ha in FY 1986, 17,613 ha in FY 1987, and 19,076 ha in FY 1988. Approximately 20,000 ha of forestland was converted to various land-use types. Percentages of total land area conversion to the following four categories for FY 1985 are: agricultural lands, 37.1% of total area converted; land for construction of public facilities, 13.8%; land for the construction of golf courses and leisure facilities, 13.6%; and other uses, 15.0%.

The percentage of converted land-use classifications changed in the late 1980s. In FY 1988, the percentage of total land area conversion for golf courses and leisure facilities increased to 42.3%, while agricultural land and public facilities decreased to 21.0% and 10.6%, respectively. At that time throughout Japan, considerable resort-area construction, including golf courses and leisure facilities, began. Many have since failed, however. In conjunction with the construction of large-scale resort areas under the Resort Area Development Law, environmental protection movements appeared. The 1991 Amendment tightened the Forest Land Development Permission System so that the environmental influence survey area was increased to include downstream areas (Goseki 1991). Pre-evaluation of flood damage possibility was added to the conditions for permission. Communication from municipality mayors related to the development of forestland and prefectural forest councils was added to the legal process to reflect local concerns.

Forestland conservation has been supported by both the Forest Protection System and the Forest Land Development Permission System. Because of failure to anticipate certain future situations, countermeasures to forest land conservation policy were needed. Namely, the Forest Land Development

Permission System was added to the forest planning system after the uncontrolled development of forestlands continued during the periods of high economic growth. Strengthening the permission system occurred after the nationwide construction of golf courses and leisure facilities became a political issue. Similarly, the problems of the Special Measures Law were already discussed in the Diet even before it was adopted. There are three main problems.

First, the Special Measures Law was established separately from the Forest Law. As the basic policy for resort area or leisure facility development plans, including forestland development, it was given high priority compared with the forest plan. To avoid duplication of examinations, the real evaluation of individual forestland development plans was omitted. Since the Japanese forest planning system places inordinate emphasis on timber production planning, functions such as controlling the development of forestland or forestland distribution within national land planning are weak. Thus, with the omission of individual inspections for forestland development, development work has become easier to plan. The forest planning system itself must be revised to add recreational components to the forest plan.

The second problem is related to differences between the Special Measures Law and the Resort Area Development Law. The concept of the Special Measures Law was to preserve forested or mountainous regions based on the sustainable forest resource management (Mizogami 1989). But there are no clear provisions relating to sustainable forest management, such as maintaining total forest area in certain planning areas.

The third problem is the lack of public hearings that would allow interested persons to speak on the issue. Since the Special Measures Law alone authorizes development plans, without any influence from local residents interested in the protection forest, legal consideration became unnecessary and thus public hearings were omitted.

The Sustainable Forest Management Rule

The concept of sustainability is basic to forest management. Japan has no rules for maintaining total forestland area either nationally or in a specific locale. The following four points highlight factors involved in this deficiency.

First, almost all Japanese administrative departments maintain a rigid administration system. There are no coordinating agencies or people in the current forest planning system. The National Land Utilization Law takes legal precedence over the 1968 Town Planning and Zoning Act (No. 100), the 1969 Agriculture Promotion Area Improvement Law (No. 58), the Forest Law, the 1957 Natural Parks Law (No. 161), and the 1972 Natural Environment Protection Law (No. 85), and plays an important role in adjusting the regional plan made by each separate law. In forest planning, the Land

Agency has hardly ever adjusted the forestland utilization plan. Laws related to regional plans are controlled by the following ministries and agencies: Town Planning and Zoning Act (Ministry of Construction), Agriculture Promotion Area Improvement Law (Ministry of Agriculture, Forestry, and Fisheries), Forest Law (Forestry Agency), and Natural Parks Law and Natural Environment Protection Law (Environment Agency). Each ministry and agency develops a national and regional plan strictly from its own perspective. In the Regional Forest Plan, for example, if a currently forested area is planned for development work (for example, construction of agricultural land and roads) in another regional plan, under the Regional Forest Plan the area is considered forest during the entire planning period. The reasons are the unclear relationships between the laws related to forest planning and the unwillingness of the ministries and agencies to cooperate.

Such problems also occurred within the Forestry Agency. For example, until the 1991 Amendment, the Regional Forest Plan for national and non-national forest lands was created by different departments without coordination. Both plans used different planning areas. The map including the Regional Forest Plan in a national forest did not even include the location of non-national forests. The 1991 Amendment required that both plans have a common planning area and that the respective agencies begin to collaborate. Even now, however, both plans are likely to be developed separately since the administrative departments creating the Regional Forest Plan are the Department of Forestry (prefectural government) for non-national forest and the Regional Forestry Office (Forestry Agency) for national forest. Neither, in fact, has much authority to transfer forests to other land uses. Accordingly, in forest planning, especially at the prefectural government level, there is nothing for the forest planner to do but assume that the currently forested area will remain so in the future. If the planner insists only on maintaining the status quo, development plans with clear economic objectives will never be influenced by the Regional Forest Plan. This has been clearly shown with forestland development during periods of high economic growth and during the latter half of the 1980s.

The second point relates to the statistical fact that Japan's total forest area has remained almost unchanged. According to the current National Land Use Plan, the total forest area was 25,100,000 ha in 1965 and 25,150,000 in 1995. It is predicted to be 25,220,000 ha in 2005. In the current plan, approximate conversion areas related to forestland from 1992 to 2005 are as follows: conversion from forest to lands for housing and factory constructions (120,000 ha), agricultural land (40,000 ha), and recreational land (30,000 ha) (Matsumoto 1996); and conversion to forest from wildlands (30,000 ha), agricultural lands (40,000 ha), and other lands, including tree planting on agricultural lands (140,000 ha). Certainly, the nationwide total forest area has remained almost unchanged statistically, but the level of forest resources has clearly changed. The percentage of artificial forests and coniferous forests, for example, has increased considerably. Forest area in and surrounding cities has clearly been decreasing. As for maintaining the total forest area, forest resource management planning must consider maintenance relating to regional units, such as regional planning areas and city and agricultural areas.

The third problem is strongly related to the first, namely, Japan's general administrative system. Many ministries and agencies have different laws and regulations that are not coordinated. Consequently, one forest area can be designated under several laws, especially forestland with non-timber values. Forests where the cutting activities are restricted by laws other than the Forest Law are listed in the 1951 Enforcement Regulations of the Forest Law (Ministry of Agriculture, Forestry, and Fisheries, Ordinance No. 54; see Table 7.3). These laws have regulations restricting harvesting based on the different objectives of each law. Other than the Forestry Agency, the ministries and agencies dealing with these laws have practically no foresters or planners or forest-protection budgets. Thus, only part of the forest resources can actually be maintained. Because of overlapping laws, forest areas are frequently under plural restrictions. Since it is difficult to remove all the restrictive regulations at once, this multiple specification seems to be an effective conservation policy. The ministries and agencies tend to lack a sense of responsibility for the forest conservation program, however, each having different opinions on protective policies such as zoning. Since there is no discussion or coordination among the ministries and agencies, the protective program functions poorly. Designating protective areas is an important policy objective in most ministries and agencies that have no forest resources staff, but after making the designation, they do not grasp the forest area's actual situation.

The fourth issue is the existence of many exceptions among forestland development work. Article 10-2 of the current Forest Law defines exceptional cases as: (1) cases where the national government or local public entities are involved in the development; (2) emergency cases due to accidents, such as fire, wind, and flood damage; and (3) cases where there is only a slight possibility of causing severe impediment to forestland conservation, but which have captured public interest and are listed by ministerial ordinance.

The rigid Japanese administrative system typically deals with exceptions when development work involves the government sector. Judging from the existence of such exceptions, the Forest Law lacks strong authority. In the third category of exceptions, the Forest Law Enforcement Regulations designate 21 excepted works, including roads, railroads, schools, museums, land improvements, land readjustments, city planning, broadcasting facilities, ports, electricity, gas, oil pipelines, and industrial water. These facilities are regulated by several laws where, in some circumstances, permission is

Table 7.3

Laws governing forest cutting activities in Japan.

Name of law	Year enacted	Law number
Erosion Control Law	1897	29
Law on Wildlife Protection and Hunting	1918	32
Fisheries Law	1949	267
Cultural Properties Protection Law	1950	214
Natural Parks Law	1957	161
Landslide Prevention Law	1958	30
Special Law on Historical Landscape Protection in Former Capital	1966	1
Town Planning and Zoning Act	1968	100
Law on Accident Protection of Steep Slope Landslide	1969	57
Forestry Seeds and Seedlings Law	1970	89
Natural Environmental Protection Law	1972	85
City Green Belt Conservation Law	1973	72
Special Measures Law on Historical Landscape Protection and Living Environment Improvement in Asuka-mura	1980	60
Endangered Wildlife Species Protection Law	1992	75

not required under the Forest Law. Railroad construction, for example, will separate the forestland from the surrounding area. School construction influences the surrounding forest area and its future management policy. These "rules by exception" ensure that each ministry or agency can practise their administrative work freely. Thus, once there is a public demand for change, in most cases the forestland is converted to another use. It is also a serious problem that the various facilities listed above are planned and constructed separately.

As these four issues relate strictly to the Japanese administrative system, it is difficult to solve forest planning system problems in isolation. Under such conditions, it is difficult to maintain forest area in the total or planning area basis and difficult to create a total forest utilization plan. Nevertheless, the Forest Protection System and Forest Land Development Permission System are the only policy programs that the administrative department of forestry can use.

Role of the Forest Planning System as an Environmental Plan

Timber management is the main objective of the conventional National or Regional Forest Plan. Certainly, it was an important policy objective in 1951, the start of the current system. However, there are 10 million ha of artificial forests in Japan, and it is necessary to develop a total forest utilization plan that includes harvesting the artificial forests that have grown during these 50 years. Under the current rigid administrative system, policy measures that forest engineers can take advantage of are strictly limited, but there are at least three important forest planning policy perspectives to consider.

First, is the environmental importance of forest resources, which comprise two-thirds of Japan's total land area. The current forest plan estimates only the standing volume, but it is necessary to clarify the composition and quantity of all forest resources, including, for example, wildlife and cultural resources. Although sustainable forest management is also discussed in Japan, it means more than just sustained timber production. The specification of indices related to sustainable forest management has been discussed in international conferences, and recently there have been some projects attempting to apply these indices to Japan's forests. It would require considerable improvement to the forest planning system to accomplish such indexing. The 1996 revised Basic Forest Plan includes the policy objective for developing forest resource data.

The second point is related to the importance of a forest resource survey. The forest resource database is managed by the departments of forestry in the prefectural governments. The database contents do not always show the real forest situation (Matsushita 1997). Trees continue to grow in the computer based on inappropriate data (Yoshida 1997). The current forest planning system is not entirely reliable even regarding timber production planning, the current system's major objective. When the forest plan is renewed to include an environmental plan, it is insufficient to just automatically calculate various figures using inappropriate data. As the first step towards renewing the forest planning system, evaluating social utilities of forest resources, and including indices for development and promotion of sustainable forest management, an accurate and practical forest resource database must be built in conjunction with a long-term continuous forest resource survey. The role of the forest resource survey in Japan has declined in general, and especially in the prefectural governments' administrative sections.

The third point is related to the role of forest ownership, especially private forest owners. When the forest plan contents are limited to timber production, private forest owners have not always felt obligated to provide accurate forest resource data because timber production is a private economic activity. From the perspective of forest resources as important environmental resources, some data, such as the distribution of important wildlife, is strictly related to the forest resource location. And, generally, the existence of such environmental resources is closely related to forest management. Private forest owners must accurately know the location, nature, and characteristics of their own forest resources, both timber and environmental resources.

Recently, the forest owners' concern for their forests has declined for many reasons, primarily the decline in timber prices and the aging of the forest owners. Subsidies for forest management, including public works in mountainous areas, have been increasing, however. With subsidy programs, forests, especially artificial forests, have grown through efforts of both the forest owner and the government. Thus, understanding and presenting the public with the real situation about forest resources are seen as part of the forest owner's obligation. Private forest owners must change their thinking to improve and broaden the current forest planning system to include an environmental perspective. In the 1996 Revised Basic Forest Plan, the existence of public use of forest resources is clearly stated. In the case of private forests, conflicts will arise between the public interest and private forest ownership. When forest owners receive subsidies in the name of public functions in their forests, however, it is necessary that they carry out forest practices to accommodate public demands. Finally, starting a new extension program for forest owners is also strongly recommended to develop their role.

References

Akai, Hideo (1984). *Shin nihon ringyo ron,* 205-208. Nippon Ringyo Chosakai, Tokyo (in Japanese).

Blandon, Peter (1996). *Igirisujin ga mita nihon ringyo no shorai* (*The future of Japanese forestry – forecasting the supply of timber*), 70-73. Tsukiji-shokan, Tokyo (in Japanese).

Forestry Agency, Conservation Division (1994). "Hoanrin seibi rinji sochiho no kaisei entyo," *Rinya-jiho* (478): 15-18 (in Japanese).

Forestry Agency (1998). *Ringyo hakusho,* fiscal 1997 ed., 44. Nihon Ringyo Kyokai, Tokyo (in Japanese).

Fukao, Seizo (1996). "Hoanrin seido no tenkai." In Handa, Ryoichi, ed., *Rinseigaku,* 2nd printing, 117-21. Bun-eido Publishing Co., Tokyo (in Japanese).

Goseki, Kazuhiro (1991). "Rinchi kaihatsu kyoka seido ni tsuite," *Kaiho* (340/341): 30-34 (in Japanese).

Handa, Ryoichi, and Sumiyoshi Ariki (1996). "Shinrin housei no seibi." In Handa, Ryoichi, ed., *Rinseigaku,* 2nd printing, 68-70. Bun-eido Shuppan, Tokyo (in Japanese).

Hashimoto, Masaki (1998). "Shinrinho tou no ichibu wo kaisei suru horitsu ni tsuite," *Kaiho* (383/384): 2-26 (in Japanese).

Kurokawa, Yasuaki (1988). "A measuring aspect of the forest planning system." In Handa, Ryoichi, ed., *Forest Policy in Japan,* 52-61. Nippon Ringyo Chosakai, Tokyo.

Matsumoto, Hiroyoshi (1996). "Dai-sanji kokudo riyo keikaku (zenkoku keikaku) no sakutei nitsuite," *Kaiho* (371): 2-10 (in Japanese).

Matsushita, Koji (1997). "Basic issues in forestry statistics in Japan." In Matsushita, Koji, ed., *A Study on the Effective Utilization of Micro-Data on Private Forest, a Report of Grant-in-aid for Scientific Research Fund by the Ministry of Education, Science and Culture of Japan,* 71-75. Kyoto University, Kyoto.

Mizogami, Kinya (1989). "Shinrin no hoken kino no zoushin ni kansuru tokubetsu-sochihou ni tsuite," *Kaiho* (327): 2-9 (in Japanese).

National Land Agency (1996). *Tochi hakusyo,* fiscal 1996 ed., 219-21. Printing Bureau, Ministry of Finance, Tokyo (in Japanese).

– (1999). *Tochi hakusyo,* fiscal 1999 ed., 272. Printing Bureau, Ministry of Finance, Tokyo (in Japanese).

Nihon Chisan Chisui Kyokai (1997). *Hoanrin Seido 100-nen shi.* 1,008 pp. (in Japanese).
Shinrin Keikaku Kenkyukai (1992). "Shinrin keikaku no kaiko to tenbo," *Kaiho* (350/351): 2-13 (in Japanese).
Tsumoto, Yorimichi (1997). "Aratana shinrin shigen ni kansuru kihon keikaku ni tsuite," *Kaiho* (75): 2-8 (in Japanese).
Yoshida, Shigejiro (1997). "Data accuracy from the point of view of forest inventory methods and up-dating methods." In Matsushita, Koji, ed., *A Study on the Effective Utilization of Micro-Data on Private Forest, a Report of Grant-in-aid for Scientific Research Fund by the Ministry of Education, Science and Culture of Japan,* 76-77. Kyoto University, Kyoto.

8
National and Regional Forest Policies
Shoji Mitsui

Objectives and Measures of Implementation in Forest Policy

Japan's forest policy has two major objectives. The first is to contribute to the stabilization of the national economy by maintaining a balance between the supply and demand of forest products, and also to ensure sufficient income and social welfare for rural people engaged in forestry. Since this objective is mainly related to the consumptive use of forests, it is not always appropriate for forest policy, unless the forestry sector is recognized as being clearly depressed and needing special support to improve the welfare of people engaged in forestry from a social policy point of view. The second objective is to maintain forests for soil and water conservation, environmental protection, recreational use, cultural values, and so on. Under the current system of private property ownership, forest policy is expected to play a significant role in optimizing such functions in the public interest.

These two objectives reflecting the functions of forests are often linked by conceptual terms such as "productive" versus "protective," "accountable" versus "unaccountable," "economic" versus "non-economic," or "private benefit" versus "public benefit." Although they appear contradictory, it should be noted that these functions are not mutually exclusive. They are, in fact, opposite sides of the same coin. Only a proportional difference exists in the social and economic context of a period, as evidenced by the history and/or geographical location of one country or region. From the period after the Second World War until 1970, for example, Japan paid more attention to the productive function of forests in pursuit of economic recovery and growth. Later, however, the protective function of forests began to be emphasized, and this trend has been steadily increasing.

Three effective measures are used to achieve these objectives: (1) direct management of forests by government authorities and administrators, (2) indirect control over forest owners through regulations, licensing systems, and so on, and (3) incentives to provide advice to forest owners through subsidies or loan systems (Handa 1990). Of these measures, discussions about

Japanese national and local forest policies focus on the third, specifically on fiscal policies typified by subsidy systems operating in the Japanese context. The following should be noted regarding current forestry loan systems:

- Wood industries rely on loans provided mainly by private finance corporations.
- Forestry management work relies on loans, up to 70% of which are provided by the Agriculture, Forestry and Fisheries Finance Corporation, a government-sponsored organization. About 90% of loans for various forest management activities are used for tree planting.

Development of Forest Policy and Public Finance

Outline of Forest Policy in Basic Forestry Law and Public Finance
The Basic Forestry Law was enacted in 1964, at the height of Japan's rapid economic growth. The objectives addressed in this law were to increase gross forest production (although that did not necessarily mean the production of timber products), improve productivity, and expand the income of forestry workers. To achieve these policy objectives, a structural change in the forestry sector was implemented in two different ways. One was for large-scale forest owners to promote a modern planning system in forest resources management, and the other was for small-scale forest owners to enlarge their management scale by promoting cooperative works.

For the first, the forest management planning system was institutionalized in 1968. This program, however, has suffered from the serious depression in the forestry sector since the importation of logs was liberalized, resulting in foreign logs steadily flooding the domestic timber market. The declining price of domestic timber made it very difficult for local timber producers to implement the forest management plan economically.

For the latter, some policy measures were undertaken, for example, to convert common forestland into individual private ownership and to introduce a profit-sharing reforestation system in government-owned national forest. The institutional framework was provided under the Common Forest Modernization Law of 1965 and the National Forest Utilization Law of 1971. It was unfortunate, however, that the enlargement of managed forestland was economically irrational, because the increase in forestry sector income could not catch up with the dramatic increase in income levels in other sectors. Domestic forestry products have stagnated due to the forestry sector depression ever since.

Under forestry structural policy, cooperation among businesses engaged in afforestation, thinning, and so on, carried out mainly by forest cooperatives, has obtained good results. It has strengthened forestry working groups in forest cooperatives. One thing that has greatly contributed to the development of forest cooperatives has been the Forestry Structure Improvement Project, established in 1964. This project has played a leading role in measures based on the Basic Forestry Law and was expected to play an important part in comprehensive enforcement of various steps to overcome small-scale forest ownership, a weak production basis, and the inadequate capital equipment.

The introduction of the Forestry Structure Improvement Project as a non-public works program affected forestry public finance in three ways: (1) the proportion of public works was reduced in the forestry budget, (2) expenditure on construction of forest roads exceeded funds for silvicultural activity in the public works allocation in the forestry budget, and (3) expenditure for the Forestry Structure Improvement Project reached 40% of funding for non-public works in the forestry budget (Murashima 1987).

In the 1970s, imported wood captured most of the Japanese timber market and began to outstrip domestic forestry production. At that time, the non-consumptive functions of forests began to be taken more seriously both in Japan and in other developed countries. Moreover, the resources of artificial forests planted over large areas after the Second World War needed a long period to reach maturity. Although at first industrial policy was the principal aim of the Basic Forestry Law, there were no objective conditions for a change from resource policy to industrial policy at that time.

Regional Forest Policy and Forestry Based on River Basin Policy
The Regional Forest Policy was introduced in 1979. The relationship between this policy and the forest policy of the Basic Forestry Law was not very clear. The aim of the Regional Forest Policy was the integrated planning of forest works and promotion of logging, processing, and distribution in each region, as set down in the Forestry White Paper of 1979. In other words, this policy aimed at the systematization of forestry production and distribution. It had the potential to rationalize the change from resource policy to industrial policy. Actually, the Forestry Enhancement Area Reorganization Plan was monitored with keen interest at the time because plans in each designated area included sections on forest work, logging, processing, and distribution. While measures for logging had scarcely been adopted before, they were enthusiastically implemented, but were spread out too thinly in 1980s.

In the first half of the 1980s, the forestry budget grew only 3% over five years because of financial reforms. Meanwhile, expansion of public works expenditure was particularly small, so spending on silvicultural activities and construction of forest roads was reduced. In non-public works, activities connected with logging and distribution increased, although expenditures on both were smaller. This may have been a reflection of the Regional Forest Policy.

In the second half of the 1980s, fiscal expansion caused by the "bubble economy" expanded the forestry budget by 21% over five years. Increases in expenditure were larger in public works than in non-public works, particularly for soil conservation activities. In non-public works, expenditure on logging and distribution decreased while expenditure on promotion of thinning and greening increased.

The Forestry Based on River Basin Policy was hammered out in the report "The Future Development Course of Forest Policy and Management Improvement of National Forest Operation," by the Forestry Administration Council (1990). To accommodate this new policy, the Forest Law was revised in 1991. The policy had two aims: (1) improvement of forests with regard to forest function, including sources of greenery and water in both private and national forests; and (2) improvement of conditions for forestry production, processing, and distribution in order to promote a new era of domestic wood.

Part of the first aim was to create forest management in relation to non-consumptive functions in accordance with public requirements, and to build a cooperative relationship between cities and towns in upstream and downstream areas in each major river basin, the same as in the forest plan area. In order to achieve this aim, the Forest Improvement Implementation Plan was drawn up. It consisted of a five-year plan for investment in public works, including silvicultural activities and forest road construction. With an appropriation of ¥3,900 billion in FY 1992-96, it became possible to guarantee expenditures in advance. Apart from this plan, however, other strategies are still vague.

Systematization of production and distribution in forestry could scarcely be implemented as a policy system in the Regional Forest Policy. The second aim was to achieve this systematization by systematic and concrete measures in order to exploit the growth of artificial forests. In each major basin, therefore, regional forest planning areas for both private and national forests were established, together with centres promoting the Forestry Based on River Basin Policy. Judging by recent examples, most areas have been slow to realize the plan except for some energetic localities such as the basin of Tosa-Reihoku in Kochi Prefecture, and the basin of Mimikawa in Miyazaki Prefecture.

In the first half of the 1990s, as the recession due to the bursting of the "bubble economy" began, public works were greatly expanded by government fiscal policy. As a result, the forestry budget has also expanded greatly since fiscal year (FY) 1992, especially in FY 1993 and FY 1995, with public works taking the lead.

Relationship between National and Local Governments
in Forest Policy

In principle, administrative works should be carried out according to Japanese law. For example, national defence and some engineering works are under the direct management of the national government. In many cases, local governments act for national government agencies as *kikan-inin-jimu*, with local authority designated by national government to carry out public affairs. In return, national government agencies distribute national government subsidies or tax grants to local governments. Thus, the relationship between the national government and local governments is epitomized by the term *"kikan-inin-jimu* and subsidies."* Government subsidies to local governments comprised 20% of about ¥71,000 billion, which was the national general account in FY 1995, and tax grants allocated to local governments comprised 19%. Thus, 39% of the national general account was allotted to local governments. On the other hand, subsidies from the national government comprise 16% of ¥83,000 billion, the annual revenue of all local governments in the Local Fiscal Program, and tax grants allocated to local governments comprise 20%. In other words, 36% of local governments' revenue is allotted by the national government.

The relationship between the national government and local governments is the same in the administration of private forests. In the national government, the Forestry Agency, an extra-departmental authority of the Ministry of Agriculture, Forestry, and Fisheries, is the agency concerned with forest policy. It is responsible for administration of private forests as well as administration and management of national forests. A large part of the administration of private forests is delegated to prefectures as the *kikan-inin-jimu*. Examples of the *kikan-inin-jimu* of forest policy include formulating the Regional Forest Plan, accepting notifications of cutting, permitting or cancelling forestland development, and designating or cancelling designation of forest reserves, based on Forest Law. In addition they permit establishment or merger of forest cooperatives according to the Forest Cooperative Law, and decree the cutting or burning of damaged trees according to Forest Pest and Disease Control Law.

The national government's forestry budget consists of the National Forest Business Special Account, the General Account (mainly for private forests), and so on. In this chapter, the General Account is examined in terms of the national forestry budget.

A large part of the national forestry budget is allocated to prefectures. A smaller part is allocated for operation of the Forestry and Forest Products Research Institute and the Forest-Tree Breeding Center, which are attached to the Forestry Agency. Another part is allocated to the National Forest Business Special Account for soil conservation works, silvicultural activities, and forest road construction. A part of the expenditure for soil conservation is allocated to prefectures. A third part is granted as a subsidy to government-sponsored organizations such as the Forest Development Corporation and the National Federation of Forest Cooperatives. The fourth part is distributed directly to municipalities. In FY 1996 the yen value of forestry funds provided by the national government to local governments was estimated to be 62% of the national forestry budget.

Prewar forest policy in Japan was very centralized because of the influence of the pivotal concept of a national forest and the primacy of the forest bureaucracy. Although this trend was not as strong after the war, the general structure of centralization in administration and public finance, described as "the *kikan-inin-jimu* and subsidies," has also characterized the relationship between the national government and local governments on forest policy. This structure, which was instituted in all government agencies after the war, is now the subject of public debate. The question concerns the nature of the relationship between the national government, local governments, and the general public with regard to forest policy.

Trends in the National Forestry Budget

The national general account in Japan was ¥77,771 billion in FY 1996. The national forestry budget was ¥589 billion, only 0.8% of the total. A survey of changes in the national forestry budget for 11 fiscal years, 1985-96, shows gradual growth in the second half of the 1980s and rapid expansion and contraction in the first half of the 1990s. Thus, if this transition is represented as an index of 100 in FY 1985, it was 121 in FY 1990, 201 in FY 1993, and 159 in FY 1996. In FY 1993, the national forestry budget reached a peak of ¥739 billion. Since then, it has fluctuated because of the fiscal policies of national government (Table 8.1).

The national forestry budget consists of public works and non-public works. Public works comprised 83% of the national forestry budget in FY 1985, rising to 85% in FY 1990 and to 86% in FY 1994 and FY 1995, before dropping to 84% in FY 1996. Public works increased by 107% to FY 1993 from FY 1985 and 63% to FY 1996 from FY 1985, whereas non-public works increased by 72% to FY 1993 and 48% to FY 1996 from FY 1985. This was the result of the national government's expansion of public works as one of a series of economic countermeasures to the economic downturn. Total expenditures on the three major forestry public works came to ¥443 billion in FY 1996: ¥262 billion for soil conservation, ¥111 billion for forest road construction, and ¥70 billion for silvicultural activities.

On the other hand, non-public works declined not only as a percentage but also in yen terms for nine years, until FY 1994; since then they have recovered a little both as a percentage and in yen terms. For example, the Forestry Structural Improvement Project account, the biggest non-public

Table 8.1

Trends in the Japanese national forestry budget, 1985-96.

Amount (millions of ¥)

Fiscal year	Total	Public works	Non-public works
1985	367,311	304,291	63,019
1986	370,466	306,844	63,622
1987	421,692	354,331	67,361
1988	420,793	354,847	65,945
1989	436,294	370,744	65,550
1990	444,848	377,043	67,806
1991	470,560	399,783	70,777
1992	544,411	463,311	81,099
1993	739,224	630,934	108,290
1994	539,404	461,518	77,886
1995	703,537	603,852	99,685
1996	588,760	495,388	93,372

Source: Forestry Agency (1987-98), Ringyo Tokei-yoran.

works account, was ¥20 billion in FY 1985, increased to ¥42 billion in FY 1993, but declined to ¥18 billion in FY 1994 before fluctuating between ¥29 billion in FY 1995 and ¥22 billion in FY 1996. In the meantime, funds transferred to the National Forest Business Special Account for retirement allowance, refunds to reduce the accumulated debts (about ¥3,800 billion at the end of FY 1997), and so on steadily increased from ¥1 billion to ¥27 billion, exceeding the amount of the Forestry Structural Improvement Project in FY 1994 and 1996 (Forestry Agency 1987-98). This indicates that the serious fiscal problem of the National Forest Business began to be a burden on public finances for private forests.

Using other data for the accounts of FY 1985-95 (Ministry of Agriculture, Forestry, and Fisheries 1990, 1995), the trend towards increasing works expenditure in the national forestry budget is as follows. First, the account of the stimulus measures for production and distribution of domestic timber increased in FY 1995. The increase in funds and the creation of new funds were especially conspicuous in the countermeasures to strengthen the wood industry and expand wood demand. Expenditure on countermeasures to maintain the forestry workforce and activate the Forestry Based on River Basin Policy also increased. It is thought that these countermeasures were aimed at strengthening the Forestry Based on River Basin Policy as a forestry production and distribution system. Second, international forest and forestry cooperation is one of the characteristics of recent forest policy. The amount spent on promotion of revegetation has increased almost constantly since FY 1990 because of the rapid increase in spending on promotion of international forestry cooperation that began that year.

In passing, new forest road consolidation works, creation of public forests, and so on began in FY 1993 under the Study Group of Forest and Upstream Villages, established by the National Land Agency, the Forestry Agency, and Ministry of Home Affairs. Prefecturally funded projects were mainly carried out under these programs, with ¥180 billion budgeted in FY 1993 and ¥235 billion in FY 1995. On the other hand, in accordance with the Specified Agricultural and Mountain Villages Law, enacted in 1993, works involving forests and mountain villages expanded as countermeasures for upstream and middle stream areas. Thus, the national forestry budget cannot be viewed only in terms of the forestry budget of the Forestry Agency.

Local Forest Policy and Forestry Budget

Long-term Views on Prefectural Forest Policy

Long-term views on forest policy, including the Long-Term Forestry Plan, the "Basic Forestry Plan," and so on, were drafted in many prefectures during the latter half of the 1960s. They were the fruit of attempts to develop an original forestry policy framework through restructuring of works delegated from the national forestry budget. In these documents, the situation and policies in each prefecture are considered. The long-term view of many prefectures around 1980 included the following goals: (1) maintenance of forest resources, (2) modernization of forestry structure, (3) promotion of the non-consumptive functions of forests, (4) stabilization of wood supply and demand and promotion of the wood industry, and (5) provision for forest workers. Nationally subsidized projects and prefecturally funded projects usually coincided with these goals (Funakoshi 1982). It is assumed that the goals of forest policy and the forestry budgets in many prefectures are currently formulated in this way.

How, then, have long-term views changed? Three cases will illustrate this:

- The "Basic Plan for Promotion of Forest/Forestry in Mie Prefecture" was published in 1995. It included the following high-priority goals: (1) establishment of low-cost forestry, (2) attracting and providing support and training for forest workers, (3) restructuring of the wood industry, (4) expansion of demand for forest products, and (5) conservation and creation of a comfortable forest environment. These goals are very similar to the five goals included in the long-term views of many prefectures around 1980. The only change was in item (1), where maintenance of forest resources and modernization of forestry structure was changed to establishment of low-cost forestry. This change was influenced by the Forestry Based on River Basin Policy. The long-term view is reflected in policies of the Forestry Agency. In other prefectures where forestry flourishes, as in Mie Prefecture, the policies of the Forestry Agency are regarded as models of appropriate long-term goals.

- On the other hand, the long-term views in some urbanized prefectures, or in prefectures with little significant forestry activity, emphasize the non-consumptive functions of forest and new trends, including public participation. For example, in the "Forest Plan of Kanagawa Prefecture" published in 1994, there were three goals: (1) conservation of forest to create highly natural landscapes, (2) creation of urban forestry to develop the multiple use of forests, and (3) interaction and symbiosis with forests.
- In the "Forest and Forestry Basic Plan of Kagawa Prefecture" published in the same year, three fundamental policies were outlined: (1) appropriate conservation and layout of forests, (2) comprehensive utilization and consolidation of forest resources, and (3) establishment of various forests through public participation.

From these examples, it can be seen that administrators in some prefectures have begun to look for original forest policies, even before the Forestry Agency breaks the convention of centralized forest policy based on public finance. It means that, little by little, practical decentralization in forest policy has begun.

Forestry Budget Trends in Prefectures

In FY 1996, gross expenditure of the forestry budget in all 47 prefectures was up to ¥1,073 billion, which amounted to 182% of the national forestry budget. Thus it is clear that prefectures play an important role in the implementation of the forestry budget.

Prefectural forestry budget expenditure gradually increased between FY 1985 and FY 1991, and rapidly expanded between FY 1993 and FY 1995. Taken as an index of 100 in FY 1985, it rose to 120 in FY 1988, 131 in FY 1991, 178 in FY 1993, 184 in FY 1995, and 175 in FY 1996. It was larger than the expansion of the national forestry budget except in FY 1993 and FY 1995.

What was the change in revenue share for the forestry budget of all prefectures in FY 1985-96? National government subsidies declined from 41% in FY 1985 to 30% in FY 1990. Subsequently, they increased to 38% in FY 1992, and then declined to 36% in FY 1993 and 34% in FY 1996. On the other hand, prefectural public loans increased from 13% in FY 1985 to 21% in FY 1987, thereby compensating for the decline in government subsidy. Further general revenue, comprising taxes and others, increased from 27% in FY 1987 to 42% in FY 1991 because of increases in tax collections due to the "bubble economy." From FY 1992, public works were expanded

to counter the recession. Government subsidies expanded rapidly and prefectural public loans also increased, while general revenue decreased. As a
result, government subsidies became 34%, prefectural public loans 30%,
and general revenue 25% in FY 1996 (Chiho Zaisei Kenkyukai 1985-96).
Thus, prefectural forestry budgets have to guarantee revenue while balancing fluctuations in the economy and national policy.

The forestry budget of Mie Prefecture is a case in point, leading to a better
understanding of prefectural forestry budgets. Changes in the indices of the
Mie Prefecture forestry budget were 100 in FY 1985, 126 in FY 1990, 140 in
FY 1992, and 179 in FY 1994. Meanwhile, the percentages of public works
changed by 66%, 73%, 70%, and 68%, respectively. These changes differ
from the national forestry budget in the first half of the 1990s. While percentages of the latter increased, those of the former decreased. Most expenditure items varied over time.

Only expenditure on forestry promotion increased consistently. The
forestry budget in FY 1994 totalled ¥17 billion, made up of 43% for land
conservation, 20% for forest road construction, 16% for forestry promotion, 9% for forestry affairs, 7% for afforestation, and 1% for expenditure
on natural conservation, forest pest and disease control, and extension campaigns of tree planting. It is notable that expenditure on forestry promotion was high, considering that it was a non-public work (Mie Prefecture
Office 1985-94).

Trends in Prefecturally Funded Projects

Prefecturally funded projects gave rise to prefectural origins of forest policy
as well as the long-term view. Most prefectural subsidies are granted as addi-
tions to national subsidies. Other prefectural subsidies of prefecturally funded
projects are independently granted, without national subsidies.

Around 1980, prefecturally funded projects consisted of small-scale afforestation, forest road construction, and land conservation, which were ineli-
gible for national subsidies, and of countermeasures for non-wood forest
products, wood processing and distribution, forest labour force, and extension campaigns of tree planting. Loan projects with prefectural financing
were for forest cooperatives, wood processing, and non-wood forest products (Funakoshi 1982).

Recent trends are not clear. Changes in volume of prefecturally funded
projects of all general construction works can be outlined. Land conservation and forest road construction are prominent, accounting for more than
70% of prefectural forestry expenditures. Nationally subsidized projects
declined from 83% in FY 1985 to 76% in FY 1994, and increased slightly to
78% in FY 1997. Meanwhile, prefecturally funded projects were 15%, 23%,
and 21%, respectively (Ministry of Home Affairs 1987-99).

In trying to understand the actual status of prefecturally funded projects, the example of the Mie Prefecture forestry budget is of interest. Percentages of prefecturally funded projects increased substantially: 13% in FY 1985, 27% in FY 1990, 32% in FY 1992, and 39% in FY 1994. Meanwhile, the yen value of prefecturally funded projects increased by 460%.

What projects led to such an increase? First, the number of small-scale projects without construction, from ¥200,000 to ¥3 million, increased. They were diverse. For example, greening and nature watching, extermination of harmful birds and mammals, preservation of pine forests, medical examination of forest workers, and aid for forest cooperatives were included. Second, land conservation, forest road work, and loan projects for housing for prefectural wood production added to the increase. Third, projects connected with the Study Group of Forest and Upstream Villages also had an influence, including improvements to forest roads and purchase of public forest (Mie Prefecture Office 1985-94). The projects involving the Study Group were different from other prefecturally funded projects, however, because spending was limited by the national government.

Municipal Forestry Budget
Regional forest policy after the Second World War had been carried out mainly by prefectures and forest cooperatives because afforestation subsidies were allocated from prefectures to cooperatives. The turning point was the 1964 Forestry Structure Improvement Project. As municipalities (cities, towns, and villages) were responsible for policy planning in this project, subsidies through municipalities increased. Subsequently, the municipalities enlarged their administrative base, through the planning and execution of the modernization project of common forests established in 1967. Moreover, the development of the Regional Forest Policy from around 1979 (for example, the Comprehensive Forest Reorganization Project) increased the status of municipalities in local forest policy.

What did the municipal forestry budget look like around 1980? It consisted of expenditures for forest road construction (34%), the Forestry Structure Improvement Project (25%), forestry promotion (16%), and so on (Suzuki 1980). Projects funded by municipal forestry budgets around 1980 consisted of grants at fixed interest rates to forest cooperatives, and grants for forest road construction, control of Japanese pine bark beetles, extension of non-wood forest products, and so on. Although these subsidies were smaller than those of the national government and prefectures, they were regarded as effective (Kamino 1982). In the mid-1980s, expenditure on forest road construction accounted for more than half the municipal forestry budget, and expenditure on municipally funded projects accounted for more than 30% (Sakasegawa 1988).

In FY 1996, the forestry budget, or gross forestry budget expenditure, of all municipalities was ¥390 billion, one and two-thirds times greater than for FY 1985 and 36% of the prefectural total. The rate of increase was 8% less than for all prefectures, indicating that the municipalities played a relatively minor role in local forest policy.

Of municipal forestry budget revenue in FY 1996, prefectural subsidies accounted for 37%, the general revenue 32%, and the municipal bond 25%. The national subsidy accounted for only 0.5%. Concerning changes since FY 1985, general revenue increased by 6% in FY 1987-91 due to the "bubble economy," but then declined as a result of the recession. Meanwhile, forest budget revenue was adjusted mainly by addition and reduction of the municipal bond. In FY 1993, the percentage of the municipal bond returned to the FY 1985 level. It then increased from 18% in FY 1993 to 25% in FY 1996 (Chiho Zaisei Kenkyukai 1985-96).

Recent trends in municipally funded forestry projects are not clear, except for changes in all general construction works concerned with forestry, which rose from 35% in FY 1985 to 42% in FY 1994 and declined to 37% in FY 1997 (Ministry of Home Affairs 1987-99). As general construction works accounted for 78% of municipal forestry budgets, it can be assumed that municipally funded projects increased until FY 1994, as did prefecturally funded projects, and then declined sharply because of the municipal fiscal crisis until 1997.

The recent status of the municipal forestry budget is not clear, and the case of Miyagawa Village in Mie Prefecture is interesting. The forestry budget of Miyagawa Village usually accounted for about 10% of its total budget. When a large-scale project like the Forestry Structure Improvement Project was introduced, however, the forestry budget temporarily increased. This is a phenomenon generally observed in other municipalities. The forestry budget of Miyagawa Village was ¥468 million in FY 1996 but rose to ¥700 million in FY 1992 and ¥809 million in FY 1993. Each year the forestry budget accounted for about 20% of the total budget. This was due to the fourth Forestry Structure Improvement Project. In FY 1992, about ¥200 million was appropriated for construction of a forestry centre, with a forest cooperative office and hall, and in FY 1993 about ¥600 million was appropriated for construction of a lumber precutting mill, where lumber such as house posts and beams are produced by computer-aided processing machines.

Shimokawa Town in Hokkaido Prefecture is a typical case of municipalities carrying out extensive municipally funded projects. Such projects in the forestry budget of this town have remained at about ¥10 million for several years, and include afforestation, cleaning and thinning, forest road construction, provision for workers, interest on management costs of a laminated-wood mill, and so on (Kanuma et al. 1996).

Issues of Decentralization and Public Participation

At present, both the welfare state that can provide for the future elderly society, and the relationship between the national government and local governments under centralized administration and public finance, are being reconsidered. The Committee for the Promotion of Decentralization, an advisory body of the Cabinet, issued an interim report on the repeal of the *kikan-inin-jimu* in March 1996. Regarding forest policy, the report stated that the authority for designating or cancelling the designations of forest reserves should be transferred from the national government to the prefectures. The national government has had direct authority for forest reserves in terms of water conservation, erosion control, and landslide prevention, and authority over other forest reserves has been delegated to prefectures as the *kikan-inin-jimu*.

Some prefectural officials reacted negatively to the idea of cancelling the designations of forest reserves, because occasionally prefectures, unlike the national government, are unable to oppose political pressure or pressure from powerful vested interests wishing to develop forests. Now the non-consumptive functions of forests are being emphasized in forest policy, and if such pressures succeed, there will be a departure from this principle of forest policy. It remains to be seen how decentralization should be carried out so that such problems are solved at the same time.

On the other hand, repeal of the *kikan-inin-jimu* of the forest planning system was not incorporated in the interim report, although it was included in the original draft. Concerning the forest planning system, public participation is being tried in Western countries, and studies have begun in Japan (Tsuchiya 1996).

As seen in this chapter, the development and implementation of national and local forest policies will be complicated by the challenge of achieving decentralization and public participation in order to reform administration and public finance.

References

Chiho Zaisei Kenkyukai (1985-96). *Chiho Zaisei Tokei Nenpo*. Chiho Zaisei Kyokai, Tokyo (in Japanese).

Forestry Administration Council (1990). *Rinsei Shingikai Toshin* titled "Kongo no rinsei no tenkai-hoko to kokuyu-rinya-jigyo no keiei-kaizen." Tokyo (in Japanese).

Forestry Agency (1987-98). *Ringyo Tokei-yoran*. Rinya Kousaikai, Tokyo (in Japanese).

Funakoshi, Shoji, ed. (1982). *Chiho-rinsei no kozo ni kansuru kenkyu*, 4-6. Chiho Rinsei Kenkyukai, Tokyo (in Japanese).

Handa, Ryoichi (1990). "Rinsei no Taikei." In Handa, R., ed., *Rinseigaku*, 97-98. Bun-eido Publishing Co., Tokyo (in Japanese).

Kamino, Shinji (1982). *Chiho-rinsei no Kadai*, 158-59. Nihon Ringyo Gijutsu Kyokai, Tokyo (in Japanese).

Kanuma, Kinzaburo, et al. (1996). "Hokkaido Shimokawa-cho ni-okeru chiiki-ringyo-kasseika no genjo to kadai" ("Current state and perspective on the development of the regional forestry in Shimokawa-cho, Hokkaido"), *Hokkaido-daigaku Nogakubu Enshurin Hokoku*

(*Research Bulletin of the Hokkaido University Forests*) 53(2): 169 (in Japanese with English summary).

Mie Prefecture Office (1985-94). *Kojiyo-to-suru Ringyo Sesaku*. Mie Prefecture Office, Tsu (in Japanese).

Ministry of Agriculture, Forestry, and Fisheries (1990-95). *Norinsuisan Yosan Kiso Shiryo*. Ministry of Agriculture, Forestry, and Fisheries, Tokyo (in Japanese).

Ministry of Home Affairs (1987-99). *Chiho Zaisei Hakusho*. Ministry of Home Affairs, Tokyo (in Japanese).

Murashima, Yoshinao (1987). "Ringyo Zaisei to Kin-yu no Tenkai Kozo." In Funakoshi, S., ed., *Chiho Rinsei to Ringyo Zaisei*, 32. Norin Tokei Kyokai, Tokyo (in Japanese).

Sakasegawa, Takegoro (1988). "Regional forest policy." In Handa, R., ed., *Forest Policy in Japan*, 365. Nippon Ringyo Chosakai, Tokyo.

Suzuki, Takashi (1980). "Shichoson ni-okeru ringyo gyosei," *Rinsei-soken Report* (11): 17. Tokyo (in Japanese).

Tsuchiya, Toshiyuki (1996). "Forester to shimin-sanka," *Ringyo Gijutsu* (654): 11-14 (in Japanese).

Part 2
Forest and Wood Products Industries in Japan

9
Logging and Log Distribution
Katsuhisa Ito

Despite the growing inventory of standing forest resources in Japan, the country has become more dependent on overseas forest resources to meet the domestic demand for logs and milled timber. For example, in 1997 wood self-sufficiency decreased to 19.6%.

It is important to strengthen the Japanese domestic log and timber distribution network by linking domestic forest resources with timber markets and marketing. Also, under the new nationwide forest management policy, "Forestry Based on River Basin," it will be more important for log production and timber processing operations in upstream and downstream areas to cooperate in conserving the basin environment, including water and flood control.

Japanese logging and log distribution, considered difficult activities, are important areas that need development and reorganization. Especially in logging, the total volume of log production and the number of workers and firms have been decreasing consistently. In recent times, very few new and younger workers have been entering the forestry workforce. New movements are being observed, however. The number of small- and middle-scale logging enterprises is increasing in the new forestry areas where artificial afforestation was widely conducted after the Second World War. In these areas, large-scale facilities are rapidly being constructed for distribution or processing of forest resources, mainly large volumes of medium-quality sugi (Japanese cedar, *Cryptomeria japonica*).

In this chapter, we examine recent trends in logging and in the log distribution system, and discuss possible future problems in both.

Log Production: The Current Situation

Background
An important role of the log marketing process is to integrate domestic forest resources with the timber processing sectors. Recently, however, the volume of log production has been gradually decreasing. There are several

reasons for this. Most of the demand is related to both foreign timber and log characteristics such as large size and homogeneity, so the share of foreign timber and logs in the domestic market is very high while the demand for domestic timber has declined. Also, profits from logging management have been decreasing. The price of domestic logs and milled timber is set by the standard market price of imported timber. In contrast, the production, distribution, and processing cost of domestic timber is expensive because of such characteristics as small and scattered forest ownership; small, inefficient, and sometimes intermittently operated sawmills; and general system inefficiency. High cost and low market price have been particularly troublesome in the logging stage.

Current Status

Log Production

Log supply and demand in Japan is shown in Figure 9.1. The volume peaked in 1973 at 119.14 million m^3 and gradually declined to 99.69 million m^3 in 1999 (84%). Domestic log production was highest in 1967 at 51.81 million m^3, declining to 18.74 million m^3 in 1999 (36.0%).

From the first half of the 1960s up to the present, log production by private and public forests (non-national forests) accounted for about 70% of total log production. Log production has also been almost dependent on private forests and production by private logging enterprises (not by the forest owners' association or direct operation of the national forest). Approximately 11,000 private logging enterprises (many of which are small-scale personal producers) and 1,100 forest owners' associations produce logs. The average annual production of a small-scale logging enterprise is 2,000 m^3; that of an average forest owners' association is 3,000 m^3.

Logs are typically produced as follows: Several people organize a log production group. They fell and limb the trees with chainsaws, and collect the whole stems with a small-log overhead wire yarding system and take them to the collection warehouse. There they are cut into logs and transported to the log market. Despite the use of high-performance forestry machines, labour productivity remains low while logging costs have remained high. There are several reasons for this. Forests for log production are separated by species, certain quality, and age classes. Forest ownership is very small and dispersed, making it difficult to collect the logs from such forests efficiently, especially under the depressed domestic forestry conditions. Also, Japan's mountainous terrain and unfavourable infrastructure is characterized by steep and widely scattered forest roads.

Japanese log production uses two types of contracting methods: by purchase and by consignment. In purchase contracting, logging enterprises

Figure 9.1

Changes in log supply and demand in Japan, 1960-99.

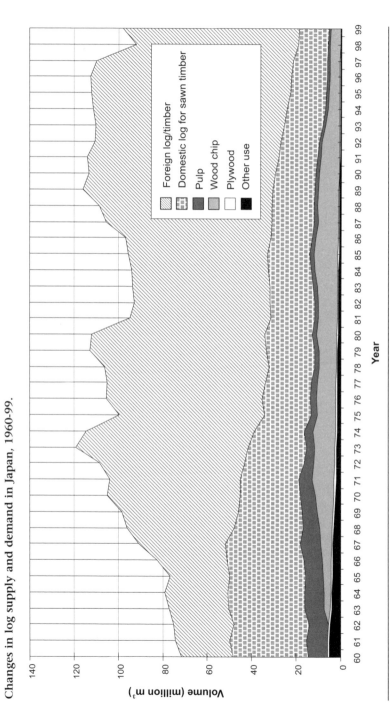

Source: Forestry Agency, "Timber Supply and Demand Report" (annual issues).

buy the trees in the forest from the owners and then start logging. In consignment contracting, logging enterprises enter into logging contracts with forest owners, with subsequent consignment of the logs for sale.

Purchase contracting is done as follows. Logging enterprises purchase standing trees from owners based on each stand's volume and estimated price (based on the market price of each stand's log volume and quality, and on logging costs). Logs are sold to log auction markets. This method entails considerable profit risk because of market price fluctuations. Overestimating log prices or underestimating logging costs can eliminate any profit. Forest owners prefer this method because they obtain money either in one lump-sum payment at the beginning of the sale or in two payments (initial and mid-sale). This method is popular among almost all logging enterprises and some forest owners' associations.

In consignment contracting, logging enterprises make a logging contract with the forest owner. Logs are cut, transported, and consigned for auction market sale. Logging enterprises then deduct their costs from sale proceeds, returning the remainder to the forest owner as income from the logs. This method is safe for logging enterprises because their costs are unaffected by log volume and quality estimates or by log market price fluctuations. Some owners dislike this method because payment is realized only after all production and consignment procedures are completed. Logging enterprises usually

Table 9.1

Number of logging enterprises in Japan, by management style.

	Number by annual logging production			
Type and nature of enterprise	50-200 m^3	200-1,000 m^3	≥1,000 m^3	Total
Personal enterprise				
Full-time	180	450	610	1,240
Part-time (logging is main)	550	1,030	970	2,550
Part-time (logging is not main)	1,200	1,040	490	2,730
Total	1,930	2,520	2,070	6,520
Private company enterprise				
Full-time	30	10	220	260
Part-time (logging is main)	10	100	600	710
Part-time (logging is not main)	330	570	1,200	2,100
Total	370	680	2,020	3,070
Forest owners' association	60	230	810	1,100
Others	30	230	170	430
Total	2,390	3,660	5,070	11,120

Source: Ministry of Agriculture, Forestry, and Fisheries (1995), *Report of Forest Trend Survey.*

pay owners between three and six months after the contract or after the start of log production. Forest owners' associations usually use this method.

Logging enterprises conduct 85% of the total share of log production from private forests, and forest owners' associations 15%. Log production from national forests is of two types: direct production by national forest employees, and contract production by private logging enterprises and forest owners' associations. Recently, as the financial condition of national forests has worsened, contract production has increased.

Table 9.1 describes logging enterprises and their management types. In 1994 there were 11,000 logging enterprises (excluding forest owners' associations) with annual production exceeding 50 m^3 each. Most (55%) small enterprises produce less than 1,000 m^3 annually. Average annual production is about 2,000 m^3 per enterprise. Sole proprietorships comprised 58% and corporations 28%.

Labour and Equipment

In recent years, the log production labour force has both aged and rapidly declined in numbers. Although common in agriculture and forestry, this phenomenon is especially acute in logging. About 1960, when economic growth began, timber demand was very strong and many rural young people joined the logging industry. Subsequently, they began a major movement from rural areas to the big cities, or from employment in agriculture and forestry to employment in other industries. Thus, the depopulation and aging of the logging labour force accelerated in rural areas because the general rural labour market in non-forestry industries expanded, because bad working conditions prevailed in logging, and because logging work required special skills and technical training (Kitagawa 1994).

The total number of both full- and part-time workers employed by logging enterprises in 1994 was 81,200, an average of 7.4 workers per enterprise. Workers averaging 60-70 years old still constitute the largest portion of the labour force and are gradually retiring. Because the entry of young workers has been observed only lately, it is estimated that the present number of workers will decrease by half in the next 10 years. Recent trends show improved working conditions in forest work. As a result of such improvement and changes in the sense of work among young people, a few young workers are gradually joining the forest workforce as they want to enjoy working with nature.

As workers age and their numbers decline, decreased log production will be inevitable, despite increasing use of high-performance forestry machines (processors, feller-bunchers, skidders, harvesters, forwarders, and so on). These are still too expensive. Also, development of the machinery system suitable for steep Japanese forest conditions is inadequate. Forest road construction is insufficient, and high-performance forestry machines are unsuitable in

Table 9.2

Logging costs for various tree species throughout Japan, 1997.

Cost (¥/m³)

Cost area	Cedar		Cypress		Pine		Larch	
	1993	1997	1993	1997	1993	1997	1993	1997
Standing tree price (a)	13,252	11,157	33,109	28,046	10,321	4,982	3,805	3,441
Logging cost								
Felling and cutting	2,782	2,792	3,619	4,206	2,890	2,985	2,371	1,996
Collecting log	3,096	3,382	4,128	4,796	3,188	2,836	1,862	1,232
Expendable supply	557	579	798	620	588	561	395	423
Depreciation	679	784	938	846	650	781	537	569
Indirect cost[1]	1,374	1,249	1,820	1,829	1,378	1,035	838	623
Total (b)	8,488	8,786	11,303	12,297	8,694	8,198	6,003	4,843
Log price at the stand (a + b)	21,740	19,943	44,412	40,343	19,015	13,180	9,808	8,284
Trucking cost (c)	2,553	2,439	2,885	2,971	2,518	2,658	2,274	2,055
Log price in market (d = a + b + c)	24,293	22,382	47,297	43,314	21,533	15,838	12,082	10,339
Logging cost/log price (%) (b/d)	34.9	39.3	23.9	28.4	40.4	51.8	49.7	46.8

1 Management cost, etc.

Source: Forestry Agency (1994, 1997), Trend and Factor Survey of the Standing Tree Price.

Japanese forests, where small and scattered stands predominate. Thus, because labour productivity remains difficult to improve, logging production costs will remain high.

The average number of logging machines per enterprise is 1.06 forest operator (one-person, track-style micro-forwarder), 1.73 yarder, and 0.09 high-performance forestry machine. Japanese forest enterprises have introduced high-performance forestry machines and combined them with the wire logging system. High-performance forestry machines in Japan are distributed as follows: 49 feller-bunchers, 162 skidders, 806 processors, 377 harvesters, 458 forwarders, and 288 tower yarders. The number of skidders and tower yarders is increasing rapidly annually (Forestry Agency 1999).

Log production costs from the latest data for 1997 (Table 9.2) reflect a survey of a private forest area (69% of all Japanese forest area) whose logging cost for indicated species exceeds logging costs in the surrounding area. This is seen as the future trend. The survey sampling method represents logging enterprises handling standing trees by commonly accepted methods for age class, volume, logging distance, logging technique, and sales contracts.

The average logging cost was ¥4,800-12,300 ($40-103) per m³, with large differences between species. The market price decreased for all species compared with 1993, but the percentage of the cost increased and reached 28-52% of the log auction market price. The logging cost in proportion to log price was generally too high to allow owners to reforest after harvest.

Wages, around 70% of the cost, were the largest portion of the expenditure. The recent trend showed a slight decline in production cost resulting from decreasing cost and increasing productivity (both of which compensated for the increasing wages) with the development of high-performance forestry machines. Productivity remains low, however, and wages continue increasing, so that the high costs have reduced logging enterprise profit and forest owners' income – a very negative situation affecting both afforestation and log production conditions. A subsidy to forest owners for reforestation is therefore necessary. Without such support, they may refrain from selling trees.

New Trends

Large-scale logging enterprises (11% of the total number) generate 56% of the total log production. Although log production is noticeably concentrated in these enterprises, their number and production volume is decreasing. In contrast, small- and medium-scale enterprises are increasing in both number and production volume (Table 9.3).

In large, afforested areas after the Second World War, especially in districts like Tohoku, Chugoku, Shikoku, and Kyushu, log production has become remarkably active (Table 9.4) due mainly to several factors: (1) timber

Table 9.3

Log volume in logging enterprise production scale, 1985 and 1994.

| Log volume | 1985 | | | | 1994 | | | |
| | Enterprise | | Production volume | | Enterprise | | Production volume | |
	Number	%	1,000 m³	%	Number	%	1,000 m³	%
50-200 m³	2,390	22	250	1	2,420	22	227	1
200-500 m³	1,880	17	570	2	2,110	19	663	3
500-1,000 m³	1,770	16	1,223	4	1,890	17	1,257	6
1,000-2,000 m³	1,770	16	2,321	8	1,680	15	2,286	10
2,000-5,000 m³	1,770	16	5,561	19	1,740	16	5,263	24
5,000 m³ and over	1,540	14	18,984	66	1,190	11	12,365	56
Total	11,120	100	28,909	100	11,030	100	22,061	100

Sources: Forestry Agency (1985), *Forestry Trend Survey*; Forestry Agency (1994), *Forestry Structure Trend Survey*; Forestry Agency (1996), *Forestry Structure Trend Survey*. Cited in Mochida, Haruyuki (1996), "The trend of logging enterprise in the 'forestry structure trend survey'," *Sanrin* (1346): 70-77.

Table 9.4

Change in number of logging enterprises in different districts in Japan, 1985 and 1994.

District	1985 Number	%	1994 Number	%	Change %
Hokkaido	770	7	570	5	-26
Tohoku	1,670	15	1,830	17	10
Hokuriku	890	8	810	7	-9
Kanto, Tozan	1,610	14	1,380	13	-14
Tokai	1,120	10	820	7	-27
Kinki	1,460	13	1,110	10	-24
Chugoku	1,270	11	1,650	15	30
Shikoku	520	5	630	6	21
Kyushu	1,810	16	2,220	20	23
Total	11,120	100	11,020	100	-1

Sources: Forestry Agency (1985), *Forestry Trend Survey;* Forestry Agency (1994), *Forestry Structure Trend Survey.* Cited in Mochida, Haruyuki (1996), "The trend of logging enterprise in the 'forestry structure trend survey'," *Sanrin* (1346): 70-77.

on areas afforested 30-40 years ago has reached harvest age and is concentrated geographically; (2) forest-tending activities such as thinning or pruning were widely conducted; and (3) a new production system, formed by young people using high-performance forestry machines and a new style of forestry operation unit, has rapidly reformed the previous production system. In the Chugoku, Shikoku, and Kyushu districts, new enterprises and cooperatives that aimed to support the development of afforestation and/ or logging workers were established, thus improving wages, welfare, and working conditions. Many such enterprises, called "The Third Sector" of business style, are established cooperatively with rural administration and private enterprises such as forestry or agricultural cooperatives. Log production with these new styles and systems will thus enhance private forest activities and may become the new forestry production system (Okada 1996).

Log Distribution: The Current Situation

Multi-stage and Other Log Distribution Systems

The log marketing system has the important role of linking the manufacturing process and domestic timber demand with logs produced in the logging sector. As required in the processing and consumption stages, the log marketing system must accomplish product distribution and feed demand information back to the logging and forest-tending processes. Recent domestic timber demand, however, has been closely coordinated with foreign timber because domestic timber has not met demand in terms of consistent quality and standard, low price, and adequate supply. The domestic log and timber marketing system thus lacks the feedback function.

The domestic sawmill log distribution channel is illustrated in Figure 9.2 (pulp and chip logs are omitted). There are three main distribution routes: (1) direct sawmill production, (2) production by loggers with shipments to sawmills, and (3) production by loggers with shipments to sawmills through log auction markets. The period from 1984 to 1991 shows slight increasing traffic through log auction markets and decreasing direct sawmill production. This resulted from the sawmills' increasing dependence on the log auction market's selective purchase of log material required by the gradually increasing mill size and specialization towards sawn timber.

There were about 13,400 Japanese sawmills in 1997, of which 43% processed specialized domestic timber, 15% processed foreign timber, and 42% processed both. Instead of specializing, domestic sawmills are small, with an average annual log consumption of about 1,900 m³ (domestic timber), 6,100 m³ (foreign timber), and 1,800 m³ (combination). In contrast, large-scale sawmills (annual log consumption of 30,000-40,000 m³) have been established in the Kyushu and Tohoku districts, where efforts to decrease cost and produce high value-added products are concentrated in afforested areas. These large-scale mills are struggling to sustain their log supplies by direct log production.

Japanese log auction marketing is a unique distribution system. Its role is to provide a smooth linkage between log production (several species harvested from small, scattered stands) and sawmills whose high-efficiency production is by timber type. Before the start of Japan's high economic growth (pre-1950), most sawmill crews did the logging (no auction market involvement) and sawn timber was subsequently processed by their respective mills. Later, to meet the rapidly increasing log demand, they specialized to increase productivity (Ogi 1994, 1995). Log auction markets have the important roles of gathering and selecting logs and acting as dispersing centres. After gathering logs, they set log prices by auction or bidding; sort logs by length, diameter, quality and species; and then transmit logs to buyers (mainly sawmills).

In the first half of the 1960s, log auction markets appeared throughout Japan except in Hokkaido, and became the most popular distribution system. Hokkaido's forests are characterized by large-scale monocultures of specialized species like larch (*Larix kaempferi*) and spruce (*Picea jezoensis*) and dispersed forestry and logging production. Log auction markets were not formed there because direct contact between log production and sawmills was easy.

Log Distribution Characteristics

Compared with foreign log distribution, domestic log distribution has its

Figure 9.2

Main distribution routes of domestic logs in Japan, 1984-91. Percentage figures are changes from 1984 to 1991.
A = volume of all log production; B = volume of all log purchasing.

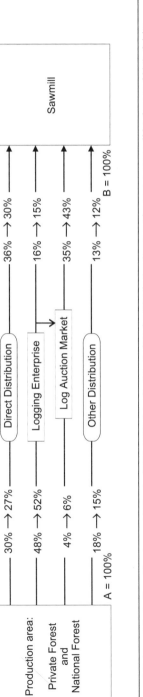

Source: Ministry of Agriculture, Forestry, and Fisheries (1984, 1991 [latest]), *Report of Timber Distribution Structure* (the next survey will be done in 2001). *Origin:* Forestry Agency (1998), *Timber Supply and Demand and the State of the Wood Industry.*

own set of characteristics that complicate its modification to improve efficiency. These characteristics include:

- a distribution system that combines small-scale and scattered forest owners with processors and that has various styles and multiple stages
- several kinds of mill and auction market distribution lots (i.e., individual sales units – *hai* – and somewhat larger aggregate consignment purchase units) that are small because of the small-scale and scattered log production operations and the predominance of small-scale sawmills
- generally small-scale enterprises conducting distribution that leads to relatively high distribution costs
- relatively inactive product development and marketing.

For domestic timber and logs used in general housing and construction, the distribution system should be larger to capture economies of scale and reduce distribution costs.

Multiple distribution stages are especially necessary for log gathering and sorting. This approach, however, negatively affects the continuous supply of low-cost, high-volume logs to consumers. As mentioned earlier, some large-scale mills or distribution centres in Kyushu and Tohoku started producing logs directly, an increasing trend. In these mills, production of good-quality, low-cost sawn timber, irrespective of log quality, is an important problem. This differs from the decreasing, but still major, Japanese processing style where most mills selectively purchase and process logs according to their customers' needs.

In contrast, the foreign log distribution system falls into two cases: (1) the import company ships directly to the specialized sawmill, and (2) shipment to the mill is through log wholesalers. Recently, activity in the first case has exceeded that in the second in Japan. Two cases also exist for sawn timber: (1) the importer ships directly to the consumer, and (2) shipment to the consumer is through timber wholesalers or retailers. Here, the percentage of the second case compared with total foreign sawn timber is increasing. The distribution of sawn foreign timber is done by affiliated companies called *Keiretsu* because the sales unit or shipment volume is large with uniform quality and piece size and because the specialized sawmill is closely connected to the import company. Thus, in contrast to the domestic side, the foreign log or sawn timber distribution system has some unique distribution characteristics and demand linkages.

Log Distribution and the Log Auction Market

Log Distribution Styles and Volume Handling Trends
Logs produced by logging enterprises are trucked to the auction market,

where they are sorted into selling units (*hai*) by species, diameter, length, quality, and shipper. Buyers create a price for each *hai* by auction or bidding, and then purchase their logs. Since auction markets generally use a consignment sale system, they collect sales commissions (5-6% of purchase amount) from buyers and sorting charges (5-6% of shipment amount) from shippers, most of whom are private logging enterprises. The log auction market is financed and managed by these charges.

In recent years, log auction market numbers have remained stable; according to the latest data, there were 402 in 1999. The total domestic log (for sawn timber) supply has decreased to 42% of the log auction market, from 31.3 million m³ in 1968 to 13.2 million m³ in 1999, but the log volume processed through log auction markets increased from 6.0 million m³ to 8.4 million m³ in the same period. Thus, the percentage of log volume handled by log auction markets has increased from 19% to 63% of the log auction market, and the average log volume handled per market has increased from 13,000 m³ to 21,000 m³ (Table 9.5). Recently, automatic log sorting machines have been used to accelerate sorting and reduce sorting costs.

Timber Use Characteristics and Log Sorting Standards
Japanese log demand stems from the need for lumber, plywood and veneer, and chips for paper. The self-sufficiency ratio has decreased to a very low level due to foreign timber substitution (Figure 9.1).

Traditional Japanese housing construction requires many different standard sizes and species for each part in the house, such as: sliding door sill (10.5 or 12.0 cm wide; 4.5 cm thick; and 2, 3, or 4 m long); post (10.5 or 12.0 cm wide; 10.5 or 12.0 cm thick; and 3 or 6 m long); and beam (10.5 or 12.0 cm wide; 18.0, 21.0, or 24.0 cm thick; 3, 4, or 6 m long). This is because each component's use is based on its characteristics, such as physical strength, which depend on the species. With length, for example, horizontal parts like sills and beams are mainly 2 m or multiples of 2 m, whereas vertical parts like columns are 3 or 6 m (there are many additional sizes besides these). Width and thickness are also classified into several conventional standards. Ornamental components used in the traditional Japanese room are classified by species, quality (knots or no knots), colour tones or combinations, and the "feeling" or style. Finally, housing parts have several distinctive variations depending on the history and nature of each area/district.

Under these complicated demand conditions and with careful selection of log quality, sawmills produce their own specialized sawn timber used mainly for housing parts. During log production, therefore, proper cutting is critical to maximize market value. The cutting strategy depends on the log's species, diameter, and quality.

Reflecting this traditional use and the buyers' behaviour, log auction market sorting standards are finely classified by quality characteristics,

Table 9.5

Overview of the log auction market in Japan, 1968-99.

Characteristic	1968	1972	1975	1980	1984	1991	1999
Number of log auction markets (a)	477	477	473	509	482	480	402
Handling volume[1] (1,000 cu. m/yr) (b)	6,010	6,260	6,020	7,700	8,050	8,780	8,370
Handling volume[1] (1,000 cu. m/market) (b/a)	13	13	13	15	17	18	21
Log production for sawn timber (1,000 cu. m) (c)	31,300	26,430	20,960	20,950	18,950	17,330	13,240
Share of log auction market (%) (b/c)	19	24	29	37	42	51	63

1 Handling volume of log auction market includes inter-market distribution volume.
Sources: Forestry Agency (1968-91), *Timber Supply-Demand Report* and *Timber Distribution Structure Report;* Forestry Agency data (1999).

including species, knots or no knots, colour tone, tree ring width, and other standards such as diameter, length, straightness, and log position (butt log, second log, etc.). Table 9.6 shows a general example of a sorting standard. Thus, the number of *hai* may reach 200-300 at each auction. Each *hai* varies from one to several stems. *Hai* of the smallest volume (1 stem, high-quality log) may range from 0.06-0.2 m³, while hai of larger volume (200-300 stems, medium-quality log) may range from 20-30 m³.

In contrast to traditional Japanese-style construction, new housing methods have appeared recently, including the "2 × 4," "wooden-panel," or "non-wooden" (timber is still used for interior ornamental parts) construction methods. Most big housing companies (over several thousand houses annually) use the new housing methods and foreign timber. Instead of domestic timber, traditional Japanese-style construction now uses more foreign timber or laminated timber for both framing and decorative parts.

In a normal, traditional-style house, the number of housing components sorted by species, size, and quality exceeds 200. Recently, to improve efficiency in distribution, processing, and precut processing, reducing the number of sizes has been encouraged (Uemura 1988). Nevertheless, houses using domestic and pure, non-laminated sawn timber processed by domestic sawmills according to the complex auction market log classification system (Table 9.6) will be limited to a small segment of overall housing demand.

Table 9.6

An example of sorting standard in the log auction market.

Sorting index[1]	Sorting kind
Species	Cypress, cedar, pine, other (larch, broad-leaved tree, etc.)
Straight/bent	Straight, bent
Log position	Butt log, second log, (over third log), top log
Knots/no knots[2]	No knots, no knots below, small and few knots, knots
Other index	
Colour tone	Clear colour, vivid colour, dark colour
Tree ring width	Tight, medium, wide, uniform, non-uniform
Size (length)	
2 m	Diameter (cm): 7-16, 18-28, over 30
3 m	Diameter (cm): 7-10, 11, 12, 13, 14-16, 18-20, 22-24, over 22 (each diameter)
4 m	Diameter (cm): 6-12, 13, 14-16, 16-20, over 22 (each diameter)
6 m	Diameter (cm): 14-16, (18-20)
Other	Special order

1 There are other sorting methods besides these, especially in length and diameter.
2 Living knots/dead knots.
Source: Personal interviews at the typical log auction market in Okayama Prefecture.

Such timber is supplied by small-scale carpenters and/or regional house builders for a traditional style in a local area.

In summary, the current log distribution system is formed by the log auction market, the major system in Japan. Each auction market's sorting standards, however, will gradually become less suitable for principal Japanese timber use because of limited volume demand.

Problems of Log Production and Distribution

Japanese standing forest volume is increasing by 70 million m^3 annually, mainly from postwar afforestation. Thus, potential supply is increasing. The demand for domestic timber remains low, however, because of the increase in foreign timber supply. This low level of demand results from high prices driven by the high afforestation and nursing investment costs of long-term management in difficult topographical conditions, by high log production and distribution costs, and by the inability to fully capture economies of scale because of the small production scale.

Under these conditions, the logging sector must achieve high productivity and lower costs, and implement a system of supplying logs rapidly and continuously to the processing sector. The recent introduction of high-performance forestry machines has occurred rapidly. Development of a distribution system capable of adapting to increasing log production volume is necessary, as is promoting low-cost production in the processing sector. Aids to such development include:

- developing the infrastructure by constructing forest roads
- developing forestry workers and enterprises, and strengthening enterprise management by enlarging firm business scale and operational area
- promoting consignment forest management that emphasizes forest owners' associations to improve small-scale forest management or logging through forest planning
- encouraging young workers to enter the forestry work force, and improving their work environment
- improving working efficiency through development and introduction of new forestry machines and technology, including high-performance forestry machines that are suited to steep Japanese forest conditions.

Achieving a low-cost supply of domestic timber requires structuring a rational log distribution system that combines the production, processing, and distribution sectors. The governmental plan of the Domestic Wood Industry Center is illustrated in Figure 9.3. There are two possible methods.

The first is through direct linkage of logging and processing, as seen in large-scale Tohoku and Kyushu district sawmills. There, it is important to

Figure 9.3

The plan of the Domestic Wood Industry Center of the production area.

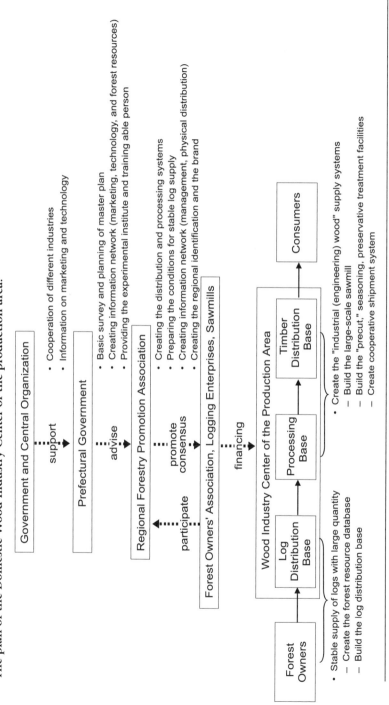

Source: Forestry Agency (1996), *Timber Supply and Demand and the State of the Wood Industry.*

continuously supply low-cost logs, process logs of all kinds and qualities without loss, and develop the sales distribution route.

Second, log auction markets are necessary as distribution centres, given the existence of small-scale and scattered forest stands and log supplies. However, if current functions such as auctioning or bidding price formation and gathering-sorting in the log auction market are required, the cost will be very high, resulting in an inefficient system. The function of the distribution centre (not the auction market) should therefore be changed to include only gathering, sorting, and accumulation of the specific log volume and quality required by the sawmill until an appointed delivery date. Log prices should be decided by the contracts between the log market and the buyer. A distribution system that limits the gathering-sorting functions is important to promote efficiency and decrease log distribution costs.

Furthermore, by developing the new nationwide forest planning approach ("Forestry Based on River Basin"), Japanese log markets will be expected to fill a greater role, as links between the domestic timber production centre and the demand sector. Thus, it will be necessary to change the market management system's gathering-sorting function to accommodate the sawmill's needs while controlling the continuity of domestic log supply and encouraging log shipment to logging enterprises and forest owners.

As mentioned earlier, the former logging production and log distribution conditions are now evolving into a new system characterized by good-quality forest resources; careful sorting, pricing, and processing; and traditional timber usage; thus, the profitability of Japanese forestry will decline. On the other hand, afforested resources are maturing rapidly and the timber demand structure is rapidly changing. A new logging and distribution system is necessary to meet both the changing afforested resource and new timber demand conditions.

References

Forestry Agency (1995). *Mokuzai Jukyu to Mokuzai Kogyo no Genkyo (Heisei Nana-nenban)*, 108-39, 150-61. Rinsan Gyosei Kenkyukai ed., Tokyo (in Japanese).

– (1998). *Mokuzai Jukyu to Mokuzai Kogyo no Genkyo (Heisei Jyu-nenban)*, 110-40, 151-73. Rinsan Gyosei Kenkyukai ed., Tokyo (in Japanese).

Kitagawa, Izumi (1994). "Sozai seisan kozo to sozaigyo no keiei kouzo." In Morita, Manabu, ed., *Rinsan Keizaigaku*, 116-29. Bun-eido Publishing Co., Tokyo (in Japanese).

Ogi, Tamutsu (1994). "Sozai ryutsu to genboku ichiuri shijo no kinou." In Morita, Manabu, ed., *Rinsan Keizaigaku*, 129-44. Bun-eido Publishing Co., Tokyo (in Japanese).

– (1995). "Sengo mokuzai shijo no ronri to ringyo." In Kitagawa, Izumi, ed., *Shinrin, Ringyo to Chusankan Chiiki Mondai*, 282-99. Nippon Ringyo Chosakai, Tokyo (in Japanese).

Okada, Shuji (1996). "Shinrin, ringyo, sanson wo meguru kyuju nendaiteki tokucho" ("The characteristics of forest, forestry, and mountainous village in the 90s"), *Ringyo Keizai Kenkyu (Journal of Forestry Economics)* 42(2): 18-27 (in Japanese).

Uemura, Takeshi (1988). *Mokuzai no Chishiki (Kaitei San-ban)*, 67-101. Toyokeizai Shinposha, Tokyo (in Japanese).

10
The Sawmill Industry
Ichiro Fujikake

Since the fuel revolution in the 1960s, which drastically reduced fuelwood consumption, the Japanese wood demand has changed so that it is now mostly from wood-based industries, especially the sawmill and pulp and paper industries. Traditional Japanese houses are built of various sizes of lumber components; the main structure of the house consists of wooden pillars and beams, and the interior and exterior are full of wood. As a result, lumber for house building accounts for a large share of Japanese wood consumption. In this chapter we will focus on the Japanese sawmill industry, which supplies much of the timber consumed in the country, and summarize some of the research results obtained so far. The first section provides an overview of the industry, while each of the following sections examines a different aspect of the industry.

Overview

Supply and Demand
We turn first to the importance of the sawmill industry in the overall wood-based industry in Japan. Table 10.1 shows the demand for logs for industrial use in Japan, and the corresponding domestic and imported supplies. Although there is also a demand for timber for non-industrial use, it is only a small percentage of industrial demand. As the data indicate, the logs for lumber consumed by the sawmill industry represent the largest share of logs for industrial use. Most of the domestic softwood and timber imported from North America and Russia is consumed by the sawmill industry; domestic hardwood is used mostly for pulp production; and tropical timber is consumed almost exclusively by the plywood industry. As for the relative significance of different wood products in the final wood consumption, after taking into account the domestic industry's timber demand and the products imported (such as lumber, chip and pulp, and plywood, but excluding paper because appropriate statistics are not available), lumber accounts for about 70% of total wood consumption.

Table 10.1

Supply and demand of logs in Japan, 1999.

	Domestic		Imported				
	Softwood	Hardwood	Tropical	North America	Russia	Others	Total
Volume[1]							
Lumber	12,751	495	522	8,458	3,731	1,492	27,449
Plywood	50	106	3,084	31	1,624	629	5,524
Chip and pulp	1,966	3,028	20	25	176	278	5,493
Others	259	82	0	19	44	6	410
Total	15,026	3,711	3,626	8,533	5,575	2,405	38,876
% by use							
Lumber	85	13	14	99	67	62	71
Plywood	0	3	85	0	29	26	14
Chip and pulp	13	82	1	0	3	12	14
Others	2	2	0	0	1	0	1
Total	100	100	100	100	100	100	100
% by log type							
Lumber	46	2	2	31	14	5	100
Plywood	1	2	56	1	29	11	100
Chip and pulp	36	55	0	0	3	5	100
Others	63	20	0	5	11	1	100
Total	39	10	9	22	14	6	100

1 In thousands of cubic metres.
Source: Ministry of Agriculture, Forestry, and Fisheries, 2000, *Survey on Lumber Production and Marketing.*

Besides the domestic forestry sector, North America and Russia are the major suppliers to the sawmill industry. The share of domestic supply is less than half, and the industry's dependence on imported timber is significant. The shares of different countries in the Japanese sawmill industry's timber supply have changed greatly since the Second World War. North American timber has almost the same share as domestic timber, and has a huge impact on the Japanese timber market and sawmill industry, which will be discussed in some detail in the next section.

Different species of timber are consumed by Japanese sawmills. Of the domestic softwood species, sugi (Japanese cedar, *Cryptomeria japonica*) has the largest share. Other species include hinoki (Japanese cypress, *Chamaecyparis obtusa*), ezomatsu (*Picea jezoensis*), todomatsu (*Abies sachalinensis*), karamatsu (*Larix kaempferi*), and matsu (*Pinus densiflora*). Douglas-fir, hemlock, spruce, fir, and pines are imported from North America, and ezomatsu, todomatsu, and karamatsu are imported from Russia. Some hardwood species such as nara (*Quercus*) and buna (*Fagus crenata*) are sawn for furniture and interior uses, but they account for a small volume compared with softwood.

Products

Table 10.2 shows the volume of lumber products by the type of use. The major use of lumber is house building, the share of which reaches 81%. As indicated in the table, lumber for housing construction is subdivided into three categories according to the shape of the cut end. Squares account for the largest volume among the three. They include pillars and beams, both of which are used to provide structural support for wooden houses. The small square is, as its name indicates, square lumber that has a small cross section in length and width. Boards are lumber with a large cross section in width. Boards and small squares are used to strengthen the pillar and beam structure as well as in various interior and exterior details. It is worth noting that there are more than 10 types of scantlings in each of these three categories that are commonly used in house building. Thus, constructing a Japanese house requires a wide variety of lumber products.

In Japanese house building, not only small squares and boards but also pillars and beams are sometimes used where their surface is visible. When lumber is used in such places, the so-called decorative lumber is preferred. Decorative lumber should have a beautiful surface. It often has a few small knots or no knots on the surface, its annual rings are equally distributed, and the distance between rings is quite small. As a result, decorative lumber has a higher value than non-decorative types. Especially when it comes to wood used for interiors of houses, the Japanese consider the appearance of the wood surface as part of the design; in fact, they rarely paint the surface,

Table 10.2

Lumber production of Japanese sawmills by use, 1999.

Type of use	Volume (1,000 m³)	%
Construction	14,666	81
Board	2,712	15
Small square	5,524	30
Square	6,430	35
Civil engineering	727	4
Box and packing	1,871	10
Furniture and fixtures	417	2
Others	484	3
Total	18,165	100

Source: Ministry of Agriculture, Forestry, and Fisheries, 2000, *Survey on Lumber Production and Marketing.*

leaving it exposed. Thus, for example, the price of precious hinoki pillars with no knots is usually several times higher than that of ordinary-grade hinoki pillars with knots. What is interesting is that the product differentiation based on the decorativeness of lumber as well as the diverse scantlings of lumber commonly used is the factor that leads to the major characteristics of the Japanese sawmill industry. We will return to this topic later.

Types of Sawmills
According to a complete survey of sawmills with a power of at least 7.5 kilowatts (kW), there were 12,240 mills in Japan in 1999, compared with approximately 25,000 around 1960. The composition of sawmills according to type of timber consumed has also changed. Of the approximately 25,000 sawmills in 1962, 75% consumed domestic timber only, 23% consumed both domestic and imported timber, and 2% consumed imported timber only. By contrast, the percentages in 1999 are 45%, 41%, and 14%, respectively.

There are two major types of sawmills in Japan: resource-location-oriented and consumption-area-oriented.

Resource-location-oriented sawmills are located in regions rich in forest resources or with easy access so that timber acquisition is easy. In the case of imported-timber mills, this type of sawmill is typically located in a factory district near the port. Resource-location-oriented sawmills usually limit the class of timber they consume, produce a limited number of products on a large scale, and distribute them to a wide area. Many consume either only domestic timber or only imported timber. In addition, because their location is determined by resource availability, such specialized sawmills often concentrate in a region to form a lumber production area. On the other hand, the management of consumption-area-oriented sawmills often stresses the distributive function by not only sawing but also purchasing lumber to respond to the house builder's requests. Although they could be large as a lumber distributor, they are typically small as a sawmill, and may use both domestic and imported timber in order to meet the customer's requests.

Although it is true that resource-location-oriented sawmills operate on a larger scale than consumption-area-oriented ones, it should be noted that there is a big difference between the average size of domestic-timber mills and that of imported-timber mills. In 1999, whereas the average annual timber purchase of domestic-timber mills was 1,800 m^3, that of imported-timber mills was 6,000 m^3. This big difference in scale between domestic-timber mills and imported-timber mills is another characteristic of the Japanese sawmill industry.

Corresponding to the two types of sawmills are two patterns of lumber distribution, as shown in Figure 10.1. As (a) in the figure indicates, products of resource-location-oriented sawmills are distributed to house builders via wholesalers and retailers. The wholesaler here may be a wholesale dealer who buys products from sawmills, or an auction market that sells products in an auction as an agent of sawmills. The consumption-area-oriented sawmill can be seen as the case in which a retailer in (a) saws timber as well as buys lumber products. The lumber distribution of this type of sawmill can therefore be illustrated as in (b).

The Postwar Development of the Sawmill Industry

The modern sawmill industry in Japan began in the Meiji era. As the free market economy spread throughout Japan after the Meiji restoration in 1868, the sawmills became concentrated in lumber-producing areas throughout the country (Handa 1986). According to Suzuki (1975), the first generation of mills with sawing machines imported from Western countries was observed around the beginning of the 20th century. Mechanization of sawmills in those days was encouraged by the increasing lumber demand caused by modernization of the economy and a series of wars. Thus the industry has a long history, but it is sufficient to go back only as far as the period after the Second World War to understand today's Japanese sawmill industry.

Postwar Recovery

The development of the Japanese sawmill industry was supported by the demand for lumber resulting from modernization and successive wars. The steady lumber market led to a situation in which any lumber products could be sold regardless of their quality or volume. Consequently, there were a large number of small mills with very primitive facilities, such as mills with

Figure 10.1

Two major patterns of lumber distribution in Japan.

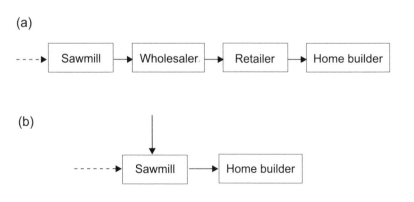

Note: The broken arrows indicate the log flow.

just a circular saw. For 10 or 20 years after the Second World War, a succession of factors – recovery from the war, a special procurement boom due to the Korean War, the beginning of rapid economic growth around 1960 – boosted lumber demand. The market remained tight, sawmill owners never worried about marketing, and small sawmills spread throughout the country.

At the same time, timber and lumber auction markets were formed, and they helped in the establishment of small sawmills that typically lacked credit availability, fundraising capacity, and marketing ability. Lumber auction markets were formed in Osaka before the war, and in Tokyo after the war; they helped small mills without their own marketing staff to sell lumber. Timber auction markets, on the other hand, date from the latter half of the 1950s. They facilitated transactions between small sawmills and small loggers who became widely established at that time, and supported sawmills and loggers by lending money on credit.

More importantly, timber auction markets induced the specialization of sawmills. Before timber auction markets existed, sawmills often purchased all logs harvested at one site, and therefore consumed a variety of logs in terms of diameter and quality. In contrast, timber auction markets sort logs by length, diameter, and quality, and sell each sorted pile of logs separately. As a result of this sorting function, sawmills were able to buy only the logs they wanted, thereby improving production efficiency. As we will see in more detail later, the combination of specialized sawmills and timber auction markets is typical of the lumber production area of today's domestic timber mills.

Period of High Economic Growth

After the establishment of many small sawmills in the tight lumber market, the Japanese sawmill industry faced a turning point when the large-scale introduction of imported timber into the domestic timber market began in 1961. Triggered by the sharp rise in timber prices in 1960 and 1961, the government set forth the Timber Import Promotion Plan in 1961, aimed at relaxing the timber market. The plan included such measures as the creation of ports for timber import and nearby industrial zones, and low-interest loans to mills that moved into the industrial zones. In 1965 there were 11 ports importing more than 300,000 m³ of timber per year; by 1973 the number had increased to 53. Thus a large volume of imported timber became accessible, reducing prices throughout the country (Murashima 1987).

As the Japanese economy grew rapidly in the 1960s, timber consumption by the sawmill industry increased by almost half, but the increase was totally supplied from abroad. From 10% in 1960, the share of imported timber in the total timber purchased by the sawmill industry reached 53% in 1970. As cheaper imported timber gradually prevailed in the timber market, not only did the market share of domestic timber decline but its total supply also decreased, because of sluggish timber prices. Domestic timber purchased by the sawmill industry remained constant in the first half of the 1960s but started to decline in 1966.

The period of high economic growth that continued until the oil crisis of 1972 was for the sawmill industry a period characterized by expansion of timber imports and by the upscaling and mechanization of mills, which was made possible by the increased supply of timber from abroad (Murashima 1987; Ogi 1994). Figure 10.2 shows the trends in total power consumption, total lumber production, and total workers in the industry, and annual production per worker. In the period of high economic growth until 1972, both total power consumption and total products increased steadily. Because the number of mills was constant through the 1960s, this meant that enhancement of capital equipment and the productive capacity of sawmills was proceeding. Annual product production per worker rose by 70% during this period.

The driving force behind the modernization of sawmills in this period was the conversion of domestic-timber sawmills into imported-timber sawmills, and the subsequent enlargement of the mills to take advantage of the rich supply of timber from abroad. Enlargement of domestic-timber mills proceeded more slowly. As imported-timber mills supplied inexpensive lumber, the lumber production areas of domestic-timber mills that were active during this period were typically those that specialized in high-quality lumber.

Figure 10.2

Trends in electrical power consumption, production volume, number of workers, and productivity in the Japanese sawmill industry, 1962-99 (index: 1962 = 100).

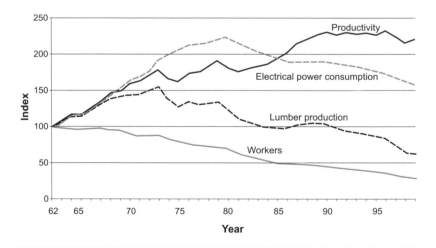

During this period, income levels rose and the demand for high-quality lumber increased. Yoshino in Nara Prefecture, the lumber production area famous for the highest-quality interior lumber in Japan (Morita 1986), and Owase in Mie Prefecture and Tono in Gifu Prefecture, famous for the hinoki decorative pillar, are good examples of the active areas. Furthermore, as lumber from imported-timber mills prevailed in the market, it became clear that the products of domestic-timber mills were often inferior in terms of product quality, including size precision and finishing. Since then, the enhancement of product quality has been one of the top priorities for domestic-timber mills. Pillar-making lumber production areas, such as Tono and Mimasaka in Okayama Prefecture, took steps to improve product quality, built a reputation for this, and succeeded in increasing their market share.

It should be noted that the development of these areas was made possible by the system of timber distribution centred around the timber auction market that developed after the war. This enabled lumber production areas to collect logs of specific diameter and quality from wider areas, and to gain a market share of particular products.

Period of Low Economic Growth

As the rate of economic growth declined after the 1972 oil crisis, lumber demand declined and the market relaxed. As we have already seen, the major demand for lumber comes from house building. This demand is influenced both by the number of building starts and by the share of wooden houses in it. Although the share of wooden houses decreased even during the period of high economic growth, because total building starts were rising, the total lumber demand from house building increased. During the period of low economic growth, however, building starts began declining and so did lumber demand. As shown in Figure 10.2, total lumber production of the industry declined during this period. The number of sawmills also started to decline around 1970, and the trend has continued to the present.

The share of imported timber continued to increase even after the economy began slowing down, and, as shown in Figure 10.2, the total power consumption of the industry also continued to grow. During the recession following the second oil crisis in 1979, however, the steady supply of imported timber, which had supported the development of the Japanese sawmill industry, broke down. The value of the yen fell against the dollar, and in the United States, the most important timber exporter to Japan, the export price of timber rose because of inflation. These factors led to an increase in the price of imported timber relative to domestic timber in Japan. Moreover, the transition from log imports to lumber imports had already begun. As a consequence, although purchases of domestic timber by the sawmill industry declined only slightly, purchases of imported timber dropped substantially. The share of imported timber in the industry's total timber purchase dropped from 63% in 1978 to 54% in 1983. The number of imported-timber mills also declined for the first time. In addition, as Figure 10.2 indicates, the rate of decline of total lumber production of the industry increased, and the total power consumption of the industry began declining too.

For domestic-timber mills, as in the preceding period, improvement of product quality, specialization, upscaling, and lowering of production costs continued to be the major issues during this period. There was no change in the general situation, but lumber production areas that specialized in decorative lumber experienced difficulties because of a drop in demand for high-quality lumber. Two reasons for this drop in demand were a decrease in the number of traditional tatami rooms that used decorative lumber, and the increased popularity of decorated glulam, a substitute for decorative lumber.

Another development during this period was that in some lumber production areas, sawmills began to tackle common problems jointly, and some of these joint enterprises were successful to a certain extent. As we saw earlier, small mills are a characteristic of the Japanese sawmill industry. Making up for this smallness by cooperation among sawmills is therefore a very important theme for the industry, and various efforts along this line have been made throughout the country up to the present.

The joint marketing of foundation boards by sawmills in Nakagawa basin in Tokushima Prefecture is a good example of this kind of cooperation among sawmills in a lumber production area. The sawmills of this area will be mentioned repeatedly in this chapter, so a description of the area is in order.

Since the Nakagawa basin is rich in forest resources and the area is close to Keihanshin Megalopolis (Kobe-Osaka-Kyoto) across the sea, the sawmill industry developed in this area from a very early period. The basin is suitable for sugi plantation, and it is famous for boards made of medium-diameter (20-28 cm in small-end diameter) sugi timber. Currently, there are about 100 sawmills consuming 340,000 m³ of timber, 60% imported and 40% domestic. Originally the area was a sugi lumber production area, but since the port for timber imports was built during the period of high economic growth, there have been both domestic-timber mills and imported-timber mills. Domestic-timber mills can be divided into several groups by main products and timber consumed. The largest group consists of sawmills producing boards of medium-diameter sugi. As the demand for boards declined because of a decrease in numbers of wooden houses and the increased use of substitutes, including plywood, about a dozen board-producing sawmill owners took notice of the foundation board used at building sites for construction work, and they tried to build a new market in the Keihanshin area by organizing a cooperative to jointly sell their products. The distribution channel for foundation board is somewhat different from that for ordinary lumber for house building. It requires sawmills to respond quickly to a large order. Joint sales through a cooperative made this possible, and sawmills in the Nakagawa basin succeeded in increasing foundation board production.

Finally, three factors concerning the trends in today's sawmill industry need to be discussed. The first is the steady advance in the conversion from timber imports to lumber imports. Lumber imports doubled during the latter half of the 1980s, and then stabilized at around 10 million m³ in the 1990s. In the late 1980s and early 1990s, the conversion was striking for imports from North America, partly due to tightening of log export prohibitions to conserve resources and protect domestic industry in Canada, and regulations affecting logs produced in national forests in the United States. More recently, lumber imports from Europe increased rapidly through the 1990s. European imports were negligible in 1990, but reached 2 million m³ in 2000. This situation will be discussed in Chapter 14.

Because imported lumber is often superior to the products of imported-timber sawmills in Japan in terms of price and quality, these sawmills are facing problems. Although the number of such mills stabilized in the latter half of the 1980s, in the 1990s it began to decline again, and the rate of decline exceeded that of domestic-timber sawmills. The remaining imported-timber sawmills try to compete through product differentiation by introducing secondary processing and responding to detailed orders from customers.

The second factor is the recent establishment throughout Japan of many large-scale domestic-timber sawmills equipped with highly productive sawing machines. The number of domestic-timber sawmills that consume more than 10,000 m³ of logs per year increased from 179 in 1991 to 234 by 1994 (Rinsan Gyosei Kenkyukai 2000). No numbers are available for after 1994, but the government *Survey on Lumber Production and Marketing* showed that the number of domestic-timber sawmills using more than 300 kW of power increased from 126 in 1990 to 174 in 1995 and 213 in 1999, whereas the total number of sawmills decreased. This trend resulted from the need for domestic-timber mills to increase productivity and cut costs in order to compete with low-priced imported lumber and glulam products.

The importance of upscaling domestic-timber mills to pursue economies of scale has long been recognized, but since stumpage prices were depressed by the low price of imported timber, the domestic harvest declined until the mid-1980s and the insufficient timber supply sometimes prevented sawmills from investing in enlarged facilities. It is interesting that although stumpage prices have not yet improved, most of the sugi plantations planted after the Second World War have already reached the age of thinning or harvesting. Consequently, the volume of the sugi harvest is increasing, especially in Kyushu, the southernmost of the main Japanese islands, where sugi grows faster than in other regions of Japan. This is supporting the upscaling of sawmills in the region. In Kyushu, where the sawmill industry responds most rapidly as the timber harvest increases, the upscaling of saw-mills will advance rapidly. In fact, the increase in the number of domestic-timber mills that consume more than 10,000 m³ of logs per year was most rapid in Kyushu, going from 35 in 1991 to 65 in 1994. Some reports claim, however, that the increased timber supply in Kyushu is a consequence of forest owners' cutting practices, and their abandoning of timber manage-ment due to stagnant timber prices. Thus, the long-term sustainability of the timber supply in this region is questionable.

The third factor is the introduction of kiln drying to sawmills. As we have already seen, an important challenge for the sawmill industry, particularly domestic-timber mills, is to improve product quality. Today's hottest issue is kiln drying, which sawmills have embarked on in response to changes in the purchasing policy of housing companies with respect to wooden components.

In order to save carpentry labour costs, housing companies now use more precut lumber. It is undesirable to precut undried lumber, however, because undried lumber components are likely to change form and size as they dry before assembly. Even after a house is built, undried lumber components are unstable and can cause problems by bending, cracking, or changing size. Anxious to avoid complaints after a house is sold, housing companies are less and less willing to use undried lumber products, and have shifted their demands to imported lumber, which is mostly kiln-dried, and glulam products. (A lumber grading system that assures the performance of lumber is another issue involving the housing and sawmill industries, although no agreement on an effective system has been reached, and few sawmills use grading machines or are certified by a grading institution.)

Kiln drying is relatively new to the Japanese sawmill industry. Kiln-dried lumber was estimated to contribute 1.5% of the total amount of lumber supplied by Japanese sawmills in 1984. As drying machines were gradually introduced to sawmills, the share of kiln-dried lumber increased to an esti-mated 8.5% in 1997 (Rinsan Gyosei Kenkyukai 2000), but this is still a small percentage. Faced with a shrinking lumber market and a prolonged eco-nomic recession, sawmills are hesitant to introduce expensive kiln-drying machines. The energy cost for kiln drying is also high, about twice as high as in Europe or North America. Furthermore, there is a technical reason for the low share of kiln-dried lumber in Japan. The water content of fresh-cut sugi, the main species used in domestic-timber mills, is high relative to that of other species, and it is technically difficult and costly to kiln-dry sugi lumber sufficiently without lowering its quality. Kiln drying is more popu-lar in sawmills cutting hinoki and imported timber than in sugi sawmills. Because sugi is the main plantation species in Japan, technical advances to address this problem are hoped for.

Domestic-Timber Sawmills versus Imported-Timber Sawmills

The differences between domestic-timber sawmills and imported-timber sawmills arise mainly from (1) differences in supply between domestic and imported timber, and (2) the consumer's special preference for domestic lumber, especially sugi and hinoki decorative lumber.

The first point that characterizes domestic-timber mills is the accumulation of sawmills that are specialized in various ways in lumber production areas. In such areas logs of various types are sorted, and then distributed to and consumed in the various specialized sawmills. Here the timber auction market typically takes care of the sorting and distribution of timber. Sawmills are specialized in terms of the type and quality of logs and lumber products to simplify production technologies and to permit different niches in one area. The diversification of mills is particularly remarkable in Japan because traditional wooden houses require many types of lumber products. Also, because of strong consumer preference for high-quality lumber, there is wide variation in the prices of lumber products with different qualities, especially decorative lumber. Different production techniques form the basis for specialization, which is influenced by two main factors: (1) because different logs are suitable for different products, production techniques vary by type and quality of timber; and (2) the production of non-decorative, ordinary lumber requires large-scale operations to enhance efficiency, whereas the production of decorative lumber requires discerning timber quality, which is time consuming.

Returning to our example of Nakagawa basin, among sugi sawmills are mills making boards of medium-diameter timber, mills making small squares of medium-diameter timber, mills making decorative pillars of large-diameter timber, mills making boards of small-diameter timber, and so on. Among the mills of decorative pillars, some specialize in very high quality pillars, whereas others specialize in average quality; some board mills make decorative boards, whereas others are devoted to ordinary board production. As a whole, all of these mills consume a variety of timber that comes from the forest resources of the region.

Another characteristic of domestic-timber mills is their small size compared with imported-timber mills. Enlargement of domestic-timber mills proceeded from 1,500 m³ of average annual timber purchase in the 1970s to 2,000 m³ in the mid-1990s, but the pace was very slow. Imported-timber mills, on the other hand, average 6,000 m³.

One of the underlying causes of this small size can be found in the historical fact that a significant number of small sawmills began operating after the war, as we have seen in the previous section. The most important reason, however, appears to be the scarcity of domestic timber, even taking into consideration the fact that mills that converted from domestic to imported timber also upscaled their operations to take advantage of the plentiful supply of imported timber.

Figure 10.3 shows the relationship between the average annual timber purchase of domestic-timber mills and the volume of softwood harvest per square kilometre of land area in 1995 for eight western Japan prefectures with sugi lumber production areas. The data show the positive relationship between the scale of the sawmills and the harvest intensity of the area. Of course, such restriction on the scale of operation imposed by timber supply is undesirable for domestic-timber mills; however, the domestic timber supply has been inactive for a long time because of the penetration of low-priced imported timber into the Japanese timber market, and this is why domestic-timber mills are, on average, approximately one-third the size of imported-timber mills. It should also be noted that because of the diversity of domestic timber supplied, each specialized sawmill could purchase a small proportion of timber supplied to the area.

As mentioned in the previous section, however, in some regions the average scale of sawmills increases in response to an increase in harvest intensity. On Miyakonojo in Kyushu, a typical example of this favourable trend, Fujiwara (1986) reported that sawmills increased their operation by adding a new line for small-diameter logs in response to the increased supply of thinnings. Unfortunately, some authors have pointed out that this increase in timber supply might not continue for long because of weakened forestry

Figure 10.3

Softwood harvest intensity and average volume of log purchases by domestic-timber sawmills in 1995 for eight western Japan prefectures.

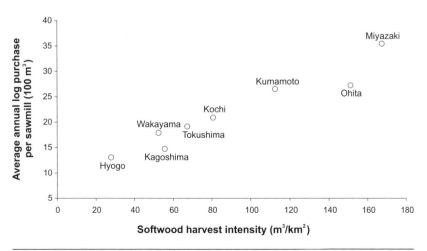

management under long-term unfavourable economic conditions for agriculture and forestry.

We have discussed the small size of domestic-timber mills, but how is it related to the production performance of the mills? A survey was conducted in Tokushima Prefecture in 1996. Figure 10.4 shows the relationship between annual timber consumption, an indication of the scale of mill, and annual timber consumption per worker, which is supposed to be a measure of productivity in quantitative terms. Sawmills are categorized into four types: sugi medium-timber mills, sugi small-timber mills, sugi large-timber mills, and imported-timber mills. For the three types other than sugi medium-timber mills, the positive relationship between mill scale and productivity is clear. It is also possible to see the same positive relationship for sugi medium-timber mills despite a relatively large dispersion of observation. These results suggest that an increase in the scale of an operation can improve production efficiency in terms of quantitative productivity.

This is not necessarily true, however, if productivity is measured in terms of value. In contrast to the quantitative productivity shown in Figure 10.4, in Figure 10.5 productivity is measured in terms of annual value added per worker. Here, the value added is defined as total sales minus total costs of logs. In this figure, the positive relationship between the scale and the productivity of mills disappears except for imported-timber mills. Even in imported-timber mills, the positive relationship is more obscure than in Figure 10.4. For the three types of sawmills that use sugi, there appears to be no relationship between the scale and the value productivity.

A reasonable explanation for these observations is as follows. In the case of domestic-timber mills, there are mills with different management strategies. Some mills pursue economies of scale by increasing the size of the operation, whereas other mills intend to add value by making high-quality products. These products are typically for decorative use and have no knots or only very small ones on the sides; they are produced using labour- and time-intensive techniques, and therefore sometimes in a relatively small operation. The productivity of the second type of mills is low in terms of quantity but high in terms of value. Among the four groups, the management strategy of pursuing high value added at the expense of quantitative efficiency is most explicit in the large-diameter sugi group, and for this group the change in distribution of the observations between Figures 10.4 and 10.5 is quite significant. Because the main product of this group is decorative pillars, the value added varies greatly according to differences in product quality, so the strategy that emphasizes product quality is particularly effective for this group. For imported-timber mills, on the other hand, because the quality of imported timber is homogeneous, and the price difference according to product quality is much less, the value productivity is closely related to the quantitative productivity.

Figure 10.4

Scale and quantitative productivity of sawmills in Japan, 1996.

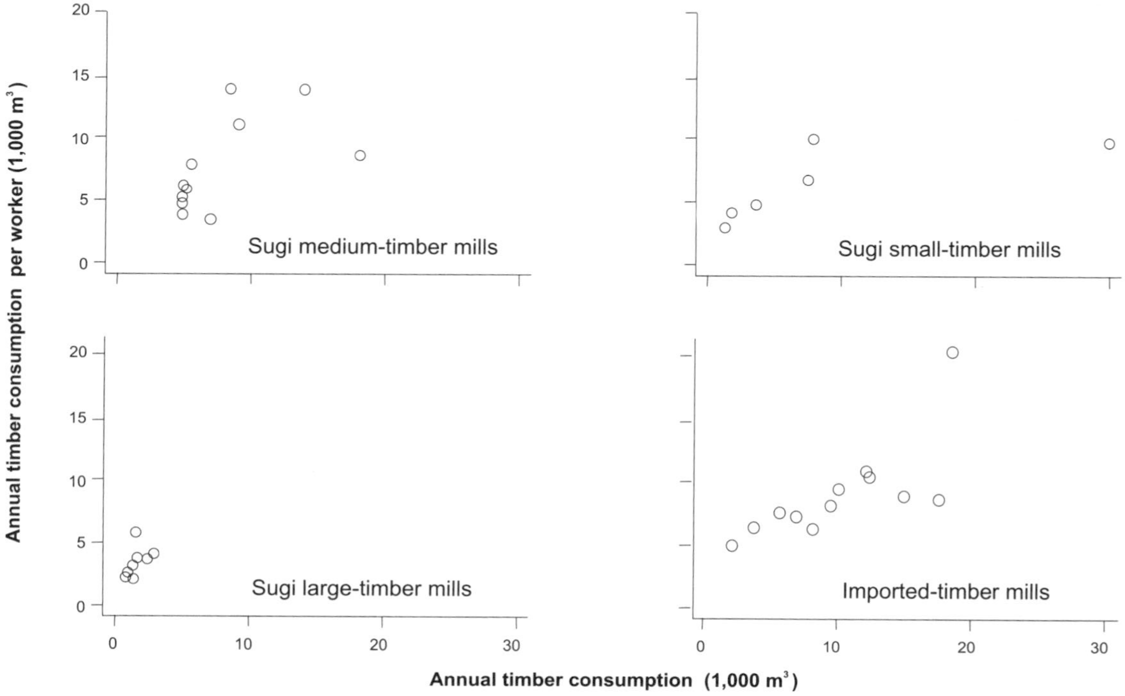

Figure 10.5

Scale and value productivity of sawmills in Japan, 1996.

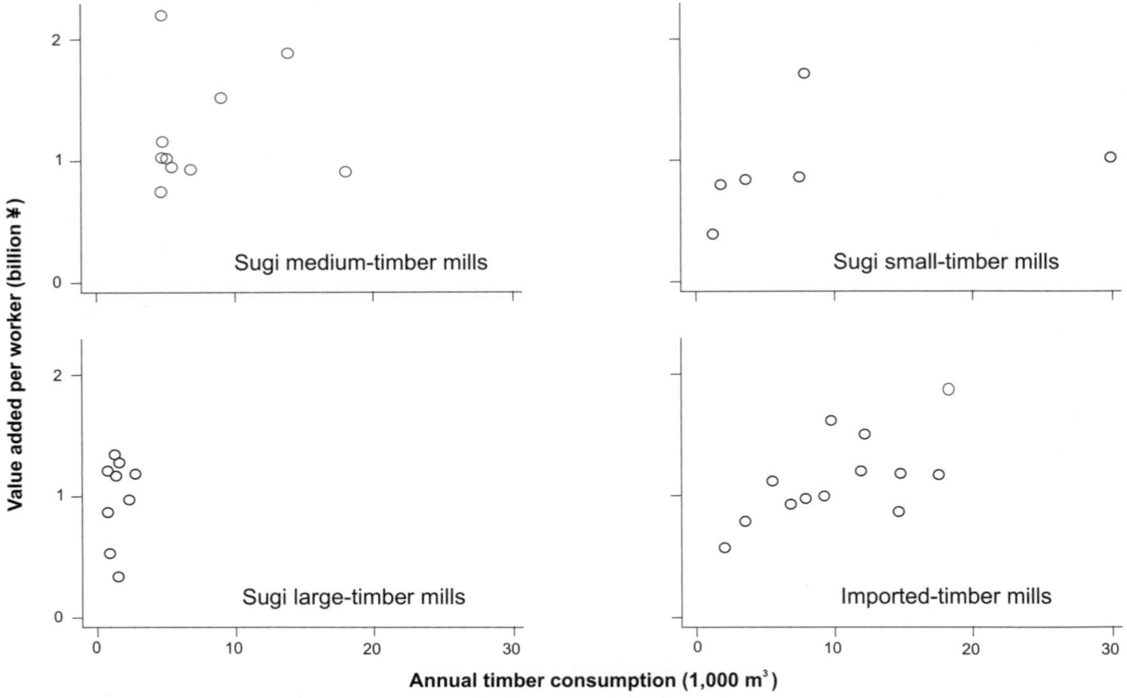

The management strategy that aims at a high value added at the expense of quantitative productivity is more effective for domestic-timber mills than for imported-timber mills for two reasons: the diversity in type and quality of domestic timber, and the big difference in the prices of lumber products made of domestic timber. Under these conditions, the small size of the mill is not necessarily a disadvantage in sawmill operation – another explanation, besides the scarcity of timber supply, for the smallness of domestic-timber mills.

Lumber Production Areas

As mentioned earlier, specialized sawmills are often located in areas where it is easy to purchase timber. Such areas sometimes have a reputation for the main products that are closely related to the forest resources available. For example, the Nakagawa basin in Tokushima Prefecture is recognized in Keihanshin district as a lumber production area specialized in sugi board production, which is supported by a large supply of medium-diameter sugi.

Since the 1960s, imported timber and lumber products have gradually prevailed in the market, and competition between lumber production areas of domestic timber has also became intense. These have resulted in differences in performance between the production areas; some areas have become sluggish, some have kept market share, others have been very active, and so on.

Under these circumstances, in almost all lumber production areas horizontal cooperation among sawmills can be more or less observed. Such cooperation was often encouraged and supported by governments, which aimed at forestry and rural development as a final objective.

The lumber production area mentioned at the beginning of this section can be considered to be based on the fact that sawmills in one area produce similar products in response to the resource quality and availability in the area. This type of lumber production area thus exists in accordance with the characteristics of the resources available in the area, regardless of the intention of each sawmill. In the 1970s, however, the creation and development of lumber production areas through horizontal cooperation among sawmills became a very important policy issue.

Horizontal cooperation works in many ways. A typical outcome of cooperation is that the exchange of information and imitation lead to equalization of production techniques and of products among sawmills; similar products from different sawmills are sold under the same brand, which is often named after the area in which the sawmills are located. In this case, attaching the same brand to products of different sawmills makes sense and is beneficial because of homogeneity of product quality based on the same production techniques, as well as homogeneity of timber resources.

The Nakagawa case is one of relatively loose cooperation. The types of cooperation in which each sawmill is more closely involved include jointly carrying out some duty related to sawmill operation so as to realize economies of scale or to gain bargaining power in the market. For example, sawmills might jointly establish a cooperative organization that sells their members' products or that purchases logs and allocates them to members' mills. Other examples include the joint use of a saw-sharpening facility, the joint marketing of bark and other scrap wood, and the joint use of a secondary processing machine or drying kiln.

The effectiveness of horizontal cooperation among sawmills varies from case to case. For instance, the effectiveness of joint marketing varies according to the kind of products sold. In the case of Nakagawa basin, the joint marketing of sugi foundation boards was successful, but the joint marketing of lumber for house building through the same cooperative was not so successful. The productive capacity to meet a large order is important where foundation boards are concerned, so joint marketing is beneficial. On the other hand, such capacity is not needed in selling ordinary construction lumber; in addition, in the case of such lumber small differences in product quality and size between products of different sawmills can greatly influence a buyer's decision.

The effectiveness of cooperation also depends on the ability of individual sawmills. For example, as member sawmills grow, the joint operation may lose a role. Soon after member sawmills start operation, when they have few customers of their own, joint marketing is critically important to members. After each sawmill has enough customers, however, and has its own brand in the market, the effectiveness of joint marketing declines.

Although horizontal cooperation is an important behaviour of sawmills in a lumber production area, it should be noted that such cooperative behaviour cannot fully cover the relationships among sawmills in one area (Kitao 1986). Needless to say, there is also competition between sawmills. Especially where equalization of production techniques and products and formation of a regional brand are concerned, competition as well as cooperation are the driving forces for sawmills to choose better techniques and imitate each other. Sawmills in one area often purchase timber from the same timber auction markets, and sometimes their customers overlap. Sawmills in one area know each other very well; as a result, their competitive relationship is very close to competition under perfect information. Thus both the cooperation and competition viewpoints are indispensable to understanding the relationship between sawmills in a lumber production area and the dynamics of that area.

To date, the major strategy in lumber production area development consists of horizontal cooperation among sawmills. Recently, however, vertical cooperation between producers, distributors, and consumers of forest products has drawn attention. Throughout the country, several cooperative enterprises between domestic-timber sawmills, house builders, architects, and, in some cases, forest owners have developed new designs for wooden houses. Such cooperation between agents with different technical information aims to add new value to the traditional wooden house. Although this type of vertical cooperation has not yet met with great success, it is noteworthy as an attempt to overcome the limitations of horizontal cooperation.

References

Fujiwara, Mitsuo (1986). "Miyakonojo no sugi nakame, shokeizai seizai." In Handa, R., ed., *Henbosuru seizai sanchi to seizaigyo*, 159-214. Nippon Ringyo Chosakai, Tokyo (in Japanese).

Handa, Ryoichi (1986). "Seizaisanchi no keisei to tenkai." In Handa, R., ed., *Henbosuru seizai sanchi to seizaigyo*, 11-24. Nippon Ringyo Chosakai, Tokyo (in Japanese).

Kitao, Kuninobu (1986). "Hokkaido no karamatsu seizai." In Handa, R., ed., *Henbosuru seizai sanchi to seizaigyo*, 255-307. Nippon Ringyo Chosakai, Tokyo (in Japanese).

Morita, Manabu (1986). "Yoshino ryoshitsuzai sanchi." In Handa, R., ed., *Henbosuru seizai sanchi to seizaigyo*, 25-44. Nippon Ringyo Chosakai, Tokyo (in Japanese).

Murashima, Yoshinao (1987). *Mokuzai sangyo no keizaigaku.* 233 pp. Nippon Ringyo Chosakai, Tokyo (in Japanese).

Ogi, Tamutsu (1994). "Seizaigyo." In Morita, M., ed., *Rinsan keizaigaku*, 53-71. Bun-eido, Tokyo (in Japanese).

Rinsan Gyosei Kenkyukai (2000). *Mokuzaijukyu to mokuzaikogyo no genkyo.* 553 pp. Rinsan Gyosei Kenkyukai, Tokyo (in Japanese).

Suzuki, Yasu (1975). "Seizai." In Nihon Ringyo Gijutsu Kyokai, ed., *Ringyogijutsushi*, vol. 5, 51-78. Nihon Ringyo Gijutsu Kyokai, Tokyo (in Japanese).

11
Home Building and the Home-Building Industry
Tamutsu Ogi

Because many residential areas in Japan were completely destroyed by bombing during the Second World War, there was a shortage of 4.20 million homes in the immediate postwar period (Ministry of Construction 1977). It was not until 1968, 23 years after the war and four years after the Tokyo Olympics, that the total number of homes first exceeded that of households, although this met the housing need only quantitatively. The number of homes constructed was only 260,000 in 1955, when high growth in the Japanese economy began. In 1960 it was 420,000. By 1965 it had increased to 840,000 homes, and in 1968 it rocketed to a record 1.20 million homes. It continued increasing, reaching 1.81 million in 1972 and a record-breaking 1.91 million in 1973 (Table 11.1). Subsequently the number fluctuated considerably, but on average 1.42 million homes were constructed annually from 1974 to 1998. By 1998 the total number of homes had reached 50 million, with a total number of households of 45 million. There were 1.12 homes per household and the proportion of unoccupied homes was 11.5%.

Table 11.2 shows the housing situation in Japan after housing supply satisfied demand quantitatively. In the case of owner-occupied homes, the number of inhabited rooms per home increased from 4.76 in 1968 to 6.02 in 1998 (26%) and total area per home increased from 97.42 m^2 to 122.74 m^2 (26%). In the case of rental homes, the increase was from 2.44 rooms to 2.84 rooms (16%), and 38.05 m^2 to 44.49 m^2 (17%). Thus, the difference in livability between owner-occupied homes and rental homes increased.

Table 11.3 shows a comparison with other countries (the United States, the United Kingdom, the former West Germany, and France). Japan is comparable to these countries in the average number of rooms, but is inferior in terms of the number of homes per 1,000 people.

Number of New Homes by Construction Method
Today wooden homes are built by three major methods of construction: the traditional post-and-beam method of construction, prefabrication of

Table 11.1

Housing completed by different construction methods in Japan, 1955-2000.

Year	Number of housing						Ratio (%)	
			Wooden				Wooden housing	Traditional housing
	Total (a)	Non-wooden	Total (b)	Traditional (c)¹	Prefabricated	2 × 4	(b)/(a) × 100	(c)/(a) × 100
1955	257,388	–	–	–	–	–	–	–
1960	424,170	–	–	–	–	–	–	–
1965	842,596	196,060	646,536	–	–	–	76.7	–
1966	856,579	214,275	642,304	–	–	–	75.0	–
1967	991,158	233,393	757,765	–	–	–	76.5	–
1968	1,201,675	315,744	885,931	–	–	–	73.7	–
1969	1,346,612	385,664	960,948	–	–	–	71.4	–
1970	1,484,556	449,056	1,035,500	–	–	–	69.8	–
1971	1,463,760	496,655	967,105	–	–	–	66.1	–
1972	1,807,581	695,735	1,111,846	–	–	–	61.5	–
1973	1,905,112	784,628	1,120,484	–	–	–	58.8	–
1974	1,316,100	446,463	869,637	842,769	26,868	–	66.1	64.0
1975	1,356,286	448,897	907,389	884,135	23,254	–	66.9	65.2
1980	1,268,626	517,973	750,653	724,619	26,034	–	59.2	57.1
1985	1,236,072	644,161	591,911	548,567	43,344	–	47.9	44.4
1986	1,364,609	730,751	633,858	581,216	52,642	–	46.4	42.6
1987	1,674,300	932,748	741,552	673,728	67,824	–	44.3	40.2
1988	1,684,644	987,377	697,267	625,620	39,185	32,462	41.4	37.1
1989	1,662,612	942,742	719,870	640,348	31,950	47,572	43.3	38.5

▼ *Table 11.1*

| Year | Number of housing | | | | | | Ratio (%) | |
| | Total (a) | Non-wooden | Wooden | | | | Wooden housing | Traditional housing |
			Total (b)	Traditional (c)	Prefabricated[1]	2 × 4	(b)/(a) × 100	(c)/(a) × 100
1990	1,707,109	979,344	727,765	642,102	34,570	51,093	42.6	37.6
1995	1,470,330	804,206	666,124	554,690	37,445	73,989	45.3	37.7
1996	1,643,266	888,970	754,296	619,028	41,575	93,693	45.9	37.7
1997	1,387,014	775,698	611,316	497,843	34,015	79,458	44.1	35.9
1998	1,198,295	653,162	545,133	447,287	29,923	67,923	45.5	37.3
1999	1,214,601	649,057	565,544	458,146	31,534	75,864	46.6	37.7
2000	1,229,843	674,029	555,814	446,259	30,341	79,114	45.2	36.3

1 Traditional = (Total wooden houses − Wooden prefabricated − 2 × 4).
Source: Economic Affairs Bureau, Ministry of Construction, *Kenchiku tokei nenpou*.

Table 11.2

Housing statistics for Japan, 1968-98.

Year	Number of inhabited rooms per house			Floor area (m²) per house		
	Total	Owned house	Rental house	Total	Owned house	Rental house
1968	3.84	4.76	2.44	73.86	97.42	38.05
1973	4.15	5.22	2.60	77.14	103.09	39.49
1978	4.52	5.65	2.79	80.28	106.16	40.64
1983	4.73	5.85	2.87	85.92	111.67	42.88
1988	4.86	6.03	2.94	89.29*	116.78	44.27
1993	4.85	6.09	2.92	91.92	122.08	45.08
1998	4.79	6.02	2.84	92.43	122.74	44.49

Source: Statistics Bureau, Prime Minister's Office, *Jyutaku tokei chosa*.

wooden homes, and the two-by-four (2 × 4) method. Until 1970 in the postwar period, over 70% of all homes were built by the traditional post-and-beam method. Accordingly, the proportion of non-wooden buildings, such as ferro-concrete and steel-frame structures, gradually increased. In 1970 the wooden building ratio (the ratio of the number of wooden housing starts to the total number of housing starts) was 70%. Because annual home construction increased until 1973, the absolute number of houses constructed by the traditional post-and-beam method also increased. During the peak of the boom in 1973, nearly 1.1 million homes were built by this method. Because of sharp fluctuations in home construction, the wooden building ratio dropped to between 40% and 49%. In 1999 only 570,000 wooden homes were built, of which 460,000 were built by the post-and-beam method. This figure is about 40% that of the peak year.

Wooden homes built by the traditional post-and-beam method have a large market share only in detached home construction. Table 11.4 shows that about 80% of detached homes are wooden. Public opinion polls conducted since 1976 (in 1980, 1986, 1989, 1993) about 80% of the respondents have indicated a preference for wooden homes. This trend can be also seen in detached home construction (Table 11.4), where the traditional post-and-beam method shows a slightly declining trend. On the other hand, the 2 × 4 method, although still small in absolute volume, has shown a marked increase. This result matches the findings of the polls that construction methods other than the traditional have become popular since the late 1980s. That is, wooden homes still enjoy overwhelming popularity (80%), but this popularity is shifting to the 2 × 4 method. Furthermore, a 1999 public opinion poll showed that the younger generations tend to avoid the traditional post-and-beam method and prefer wooden home construction by other

Table 11.3

Comparison of international housing data from the period 1984-91.

Country	Population[1]	Number of houses[2]	Houses per thousand people	Average rooms per house	Average persons per room	% vacant houses	% owned houses
USA	24,808 ('89)	105,729 ('89)	427 ('89)	5.3 ('89)	0.40 ('89)	11.3 ('89)	64.0 ('89)
UK	5,706 ('88)	22,749 ('89)	400 ('89)	4.9 ('89)	0.50 ('88)	5.4 ('91)	67.6 ('91)
Former West Germany	6,199 ('89)	26,280 ('87)	430 ('87)	4.4 ('87)	0.50 ('87)	2.7 ('87)	39.3 ('87)
France	5,661 ('90)	26,463 ('90)	464 ('90)	4.4 ('90)	0.70 ('84)	17.2 ('88)	51.2 ('84)
Japan	12,361 ('90)	42,007 ('88)	342 ('88)	4.9 ('88)	0.66 ('88)	9.4 ('88)	61.3 ('88)

1 In tens of thousands.
2 In thousands.
Sources: United Nations, *Annual Report of Housing and Building Statistics for Europe* (UK, West Germany, France); Housing Bureau, Ministry of Construction, *American Housing Statistics, Housing and Construction Statistics, Labor Force Surveys,* etc. (USA); Statistics Bureau, Prime Minister's Office, *Housing Statistics Survey* (fiscal 1988), *National Census* (fiscal 1990), and *Monthly Report of Population Estimate* (Japan).

Table 11.4

Detached houses completed by different construction methods in Japan, 1988-99.

Year	Number of houses						Ratio (%)	
			Wooden				Wooden housing $(b)/(a) \times 100$	Traditional housing $(c)/(a) \times 100$
	Total (a)	Non-wooden	Total (b)	Traditional (c)	Prefabricated	2×4		
1988	650,342	123,816	526,526	–	–	–	81.0	–
1989	648,059	123,223	524,836	472,186	21,363	31,287	81.0	72.9
1990	637,251	124,126	513,125	456,574	22,769	33,782	80.5	71.6
1991	566,221	108,646	457,575	405,131	22,228	30,216	80.8	71.5
1992	592,718	110,405	482,313	428,865	22,542	30,906	81.4	72.4
1993	663,161	122,090	541,071	479,724	25,882	35,465	81.6	72.3
1994	728,017	134,020	593,997	518,460	29,907	45,630	81.6	71.2
1995	686,265	138,686	547,579	464,266	30,932	52,381	79.8	67.7
1996	800,820	165,094	635,726	530,415	35,485	69,826	79.4	66.2
1997	629,412	126,180	503,232	417,434	28,099	57,699	80.0	66.3
1998	556,529	109,627	446,902	372,507	24,430	49,965	80.3	66.9
1999	599,059	116,160	482,899	398,892	26,607	57,400	80.6	66.6

Source: Economic Affairs Bureau, Ministry of Construction, *Kenchiku tokei nenpou*.

methods, or even non-wooden houses. The decline of the traditional method may also be accelerating in detached home construction, where it has long enjoyed an unchallenged position.

Table 11.5 shows the share of each construction method in 1999, with the traditional post-and-beam method at 38%, prefabricated houses at 15%, the 2 × 4 method at 6%, and others at 41%. The traditional post-and-beam method is defined as building a structure in which posts are erected and joined by a foundation consisting of lower horizontal members, and beams and crossbeams consisting of upper horizontal members, with a roof. Wooden members used in this method fall into three groups:

- Group I consists of structural materials such as foundations, posts, beams, and crossbeams. These materials require large cross-sectional members that constitute the structural units of a home.
- Group II consists of groundwork materials that have a role supplementary to that of structural materials and a smaller cross section. These are "in and out of sight" materials except for the structural materials.
- Group III is for fixture materials. These may be compared with materials required to make Japanese-style rooms. Materials in this group are much more expensive than those in the other two groups. Fancy domestic lumber has traditionally been used for fixture materials.

Imported timber overwhelms domestic timber in the share of house-building materials. Domestic lumber is mostly used as members for Japanese-style rooms in the traditional method of construction, that is, the Group III fixture materials and a very small portion of post materials. The fact that people now desire fewer Japanese-style rooms is putting pressure on the domestic-timber industry.

Imported lumber is used very extensively in the traditional post-and-beam method of construction. As will be described later, domestic lumber is rarely used in other construction methods. Homes built by other construction methods have Japanese-style rooms in which fixture materials are used, but almost all such materials are being replaced by laminated wood. In laminated

Table 11.5

Houses completed by different construction methods in Japan, 1999.

	Total	Traditional[1]	Prefabricated	2 × 4	Others[2]
Number	1,214,619	458,146	185,724	75,864	494,885
% of total	100.0	37.7	15.3	6.2	40.7

1 Traditional = (Total wooden houses − Wooden prefabricated − 2 × 4).
2 Others = (Total − Traditional − Prefabricated − 2 × 4).
Source: Economic Affairs Bureau, Ministry of Construction, *Kenchiku tokei nenpou.*

wood, fancy veneer from domestic lumber is overlaid on imported lumber. The only domestic lumber used in non-traditional methods of construction is sometimes a thin veneer of surface material of laminated wood for interior finishing.

Prefabricated home construction is defined as "a method of construction in which important structural members of building such as walls, posts, floors, beams, roofs, stairs, etc., are produced on a massive scale in a factory, and these materials are used to erect buildings at sites" (Ministry of Construction 1999). These are categorized as wood, steel-frame, and concrete prefabricated methods according to the type of structural materials employed. This method first appeared on a commercial basis in the 1960s. It did not reach 10% market share until 1980. At present, it accounts for 15% of total housing starts (Figure 11.1).

The 2 × 4 construction method was introduced from North America. The full-scale application of this method started only in 1974, but it soon surpassed the share of wooden prefabricated homes. The remaining construction methods include all non-wooden types of homes that are not produced in factories, such as steel-frame construction, reinforced-concrete, and steel-frame construction/reinforced-concrete.

A fusion or combination of construction methods has been encouraged recently. For example, the "traditional + panel" method combines the traditional and panel (a type of prefabricated home method) methods, and

Figure 11.1

Ratio of prefabricated houses to total home construction in Japan, 1980-99.

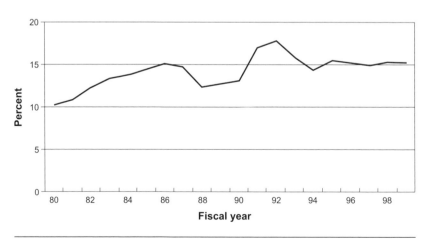

Note: Two-by-four houses are included up to fiscal 1994.
Source: Ministry of Construction (1999).

the "two-by-four + unit" method combines the 2 × 4 method and the unit method (another type of prefabricated home method). These phenomena were trends among home-manufacturing companies in which prefabrication of homes was encouraged, and they appear to have further accelerated prefabrication.

In the past, the external appearance of a house signalled the method of construction. Today, neither exterior nor interior appearance of homes is a good indicator of the construction method. From visual observation, it is difficult to differentiate construction methods. Whatever the methods used, most homes today are built using a Japanized Western style, and Japanese-style homes are gradually declining.

The volume of wood used per detached home differs among methods as follows: traditional method, 0.179 m^3/m^2; 2 × 4 method, 0.221 m^3/m^2; prefabrication of wooden house, 0.153 m^3/m^2. The 2 × 4 method consumes the most lumber per square metre, followed by the traditional method and then the prefabricated wooden home (Sakai et al. 1991).

Apartments

The basic Japanese home before the war was a detached house. In addition, there were rented row houses, called *nagaya*, which incorporated two or more units separated by a partition wall. This imitated the detached house and the two-story building. After the Second World War, a great number of tall and medium-height (three stories or more) multiple-family dwellings were built, mainly in big cities. In Japan, these dwellings are called "apartments."

According to the Ministry of Construction, an "apartment" is rather narrowly defined as a house and lot (subdivided from a larger lot) offered for sale with characteristics of "common structure" ("common structure" means that "two or more housing units exist in a building, where some or all of the halls, corridors and stairs are commonly used" [Ministry of Construction 1996]) and with either ferro-concrete and steel-frame, or ferro-concrete, or steel-frame structure.

In general, "apartment" refers to a tall (three or more stories) or medium-height multiple-family dwelling, including a house and lot offered for sale and a rented house. With industry and population overly concentrated in urban areas due to the rapid economic growth beginning in the mid-1950s, the apartment emerged as intensive land use type of structure for cities where the land price was very high. In the beginning, therefore, most apartments were sold by targeting a limited number of large-income earners who could afford such dwellings. Since the mid-1960s, apartments were developed in less expensive suburban areas, where they became popular with the general public, and eventually became a common type of residence in cities. The number of apartments constructed in 1999 was 192,000: 100,000 in the Metropolitan area (Tokyo, Kanagawa, Chiba, and Saitama) (52.1%), 43,000 in Kinki area (22.4%), and 13,000 in Chubu area (6.8%). These three major areas accounted for 81.2% of the total number of newly constructed apartments in Japan in 1999, indicating that apartments are typical urban housing units (Ministry of Construction 2000).

Imported Houses

Imported houses are housing units that are imported from other countries as materials or in a kit form and assembled in Japan (Imported Houses Promotion Association 1994). The supply of imported houses began to increase in the mid-1980s, stagnating in the early 1990s before suddenly increasing again in 1994. In 1998 the annual construction of imported houses was 7,515 units, which was 17 times that of fiscal 1985 (423 units) (Table 11.6).

In 1995 an estimated 77 companies handled imported houses from 44 foreign companies: 26 American, 4 Canadian, 2 Danish, and 2 Norwegian. Two-by-four homes accounted for 75% of all imported houses. Some 86% are used as residences and 10% as resort villas. The price per *tsubo* (about 3.3 m²) is ¥500,000-800,000 (Ministry of Construction 1995b). The floor space is 30-60 *tsubo* (about 99-198 m²) (Imported Houses Promotion Association 1994).

Characteristics of Traditional Post-and-Beam Japanese Houses

Builders Using the Traditional Post-and-Beam Construction Method

Most traditional post-and-beam houses are built by small- to medium-sized carpenters or *komuten* (builders). About 616,000 such homes were built in 1994; 38,000 were built by 15 builders who constructed more than 500 homes a year, and 10,000 were built by 41 builders who constructed 100-500 (exclusive) houses a year. Together these two types of builders construct 48,000 houses a year (7.8%) (Ministry of Construction 1995a; Nikkei Sangyo Shouhi Kenkyujo 1996). This shows that over 90% of traditional Japanese homes are built by builders who build fewer than 100 homes a year. According to the *Fiscal 1991 Business Statistics Report* issued by the Bureau of Statistics of the Prime Minister's Office, the number of carpenters and *komuten* in 1991 was 155,000, with an average of 4.4 employees each. The traditional method of home building is thus carried out by a number of very small businesses.

Since the 1980s, the number of carpenters and *komuten* has decreased. This trend is likely to continue. The decline in the number of carpenters has contributed to the growth of the precut method (see below).

According to the above-mentioned report, the average number of employees per carpentry business is 2.9. Thus, a carpenter builds several homes annually employing several staff members, and it is not necessary to have an office for them. Before the emergence of large-scale home manufacturers, all ordinary homes in Japan were built by these carpenters. Companies

Table 11.6

Supply of imported houses in Japan, 1985-98.

Year	Number of imported houses						Number of houses imported by Sweden House[3]	Number of enterprises
	Log house	2 × 4	2 × 4, 2 × 6, panel	Framework	Others	Total		
1985	–	–	–	–	–	423	41	–
1986	–	–	–	–	–	610	80	–
1987	–	–	–	–	–	877	138	–
1988	–	–	–	–	–	1,078	250	–
1989	188	425	128	107	521	1,369	350	43
1990	213	500	147	111	515	1,486	380	54
1991	174	582	206	115	336	1,413	395	62
1992	181	579	208	143	340	1,451	420	66
1993	215	540	208	189	413	1,565	387	65
1994[1]	407	975	1,068	106	63	2,619	552	71
1995[1,2]	466	1,903	1,565	439	202	4,575	810	77
1998[1]	–	–	–	–	–	7,515	–	–

1 Fiscal year.
2 Estimated number.
3 Figures were acquired by hearing.

Sources: House Product Promotion Foundation (1991), Survey Report on Improvement of Housing Supply System to Cope with Internationalization (1985-88); Imported Houses Promotion Association (1994), Report of Imported Houses Promotion Association (1989-93); Ministry of Construction (1995), Questionnaire on Imported Houses (1994-95); Forestry Agency (2000), Ringyo Hakusho (1998).

that are slightly larger in size than carpenter's businesses and that have sales offices are called *komuten*. Their average number of employees is 5.4. Both carpenters and *komuten* depend solely on personal connections for their business. As a result, both the scale of their operation and their volume of work are very limited. They also handle non-traditional construction methods to meet their clients' demand. In particular, 2 × 4 homes are gaining popularity now, leading to an increase in the number of carpenters building such homes. This trend will attract a great deal of attention in the future.

Precut Method
In the traditional post-and-beam method of construction, it was common practice for carpenters to manually cut members for the framework at construction sites or at their workshops. This process required much technical skill and much time. "Precutting" replaces this on-site cutting process with machining in a factory.

Precut factories first appeared in the mid-1970s and recorded sudden growth from the late 1980s to 1990. Over the 14 years between 1985 and 1999, the number of precut factories increased sixfold, from 142 to 888 (Table 11.7). According to the National Wooden Home Precut Association, 45% of traditional Japanese homes were built by the precut method in fiscal 1998.

Table 11.7

Precut factories in Japan, 1985-99.

Year	Number of precut factories	Increase ratio[1]
1985	142	–
1986	161	13.4
1987	215	33.5
1988	284	32.1
1989	387	36.3
1990	483	24.8
1991	589	21.9
1992	600	1.9
1993	654	9.0
1994	717	9.6
1995	784	9.3
1996	874	11.5
1997	881	0.8
1998	888	0.8
1999	888	0.0

1 Increase ratios compared with the previous years.
Source: National Wooden Home Machine Precut Association.

From 1987 to 1990, Japan enjoyed an unprecedented housing boom, with a record 6.73 million homes being constructed during this period. The rapid increase in the number of precut factories was triggered by the extreme shortage of carpenters and artisans during this housing boom. This is clearly reflected in the rate of increase of precut factories shown in Table 11.7.

There are two types of precut factories: those affiliated with a home manufacturer and independent factories. The former includes various companies, from substantial subsidiaries financed by a home manufacturer to exclusive subcontractors. This type of precut factory engages exclusively in precut machining for a particular home manufacturer. On the other hand, independent precut factories have no such affiliation with home manufacturers, and have to find their own markets. Small- to medium-sized carpenters and *komuten* use these independent precut factories.

One of the most significant changes resulting from the emergence of precut factories was that, as the technology has developed, precut products have contributed greatly to improving the quality of components. Many precut factories have introduced the CAD/CAM (computer-aided design/computer-aided manufacturing) system, in which a plan is drawn up by computer-aided equipment and automatically read by computer, and materials are cut exactly to plan. To make production flow smoothly, precise dimensions within a certain permissible range are required. To this end, dried lumber must be given priority. Unfortunately, only a limited number of sawmills in Japan can afford to manufacture lumber that meets this requirement. For example, lumber produced in Yoshino, Nara Prefecture, which is the most famous lumber-producing centre in Japan, was examined. In February 1995, a survey of the dimensions of hinoki square pillars from 13 local sawmills showed that only 5% met the requirements of precut factories (Nara Prefecture 1996).

For precut factories to have efficient production and to fully realize capacity, lumber must have a certain dimensional accuracy; that is, it must have the necessary quality to allow precut factories to grow. In Japan, where green lumber is generally accepted and quality control is low, improvement in domestic lumber does not seem easily achievable. For this reason, precut factories tend to use only imported lumber or lumber from major domestic manufacturers. Recently, a large amount of glue-laminated wood has been used for precut components. Glue-laminated wood is used not only for structural members such as posts, beams, crossbeams, and foundations but also as spacers. The use of glue-laminated wood has become markedly popular since 1995, when a major earthquake in January caused great disaster in and around Kobe. This made ordinary people realize the importance of the physical properties of materials. Laminated wood is much better than lumber in dimensional accuracy, homogeneous quality, and strength. In the numerically controlled precut production line, laminated wood is better by far than lumber as a raw material.

Home Manufacturers

In Japan, the massive influx of population into urban areas due to the high economic growth beginning in the mid-1950s created an enormous demand for housing, especially in expanding urban areas. Traditional carpenters and *komuten* solely depending upon relatives and people in the local community were not able to cope with this massive demand. Under such circumstances, home manufacturers able to commercially supply homes on a massive scale emerged and gradually took root in the community. One of their major characteristics is that they manufacture and sell homes as commercial products. Because it is necessary for them to constantly provide potential customers with information about their products, home exhibition centres have been established. The exhibition of homes at such centres provides a guide to home manufacturers.

Home manufacturers advertise and carry out business at their exhibition centres by providing visitors with information on "hardware," such as method of construction and structural characteristics, and "software," such as how to maximize living conditions. Together with housing magazines, these home exhibition centres are now the most important information resource for consumers.

Home manufacturers enjoy a 25% share of the housing market, although this varies from year to year. In the Japanese housing market, they exert the most influence in every aspect of housing, from type of house to interior design.

A model house in a home exhibition centre is used as a sales office, where contracts with home buyers are usually signed. The number of such centres nationwide was 332 in 1994 (not including the number of single model homes displayed by individual companies). Sekisui Prefab Homes, Ltd., the largest home manufacturer, exhibits 353 model homes in 332 home exhibition centres (Yano Keizai Kenkyujo 1994), the largest of which, in Yokohama, exhibits more than 100 model homes. One centre in Tokyo and one in Osaka exhibit more than 50 model homes each.

Table 11.8 shows the number of homes built by home manufacturers in fiscal 1994. The top 10 companies and the top 20 companies enjoy a 17% and 19% share, respectively. The top 59 companies, each of which built 500 or more homes a year, accounted for a 22% share. Companies that build 500 or more homes a year are considered to be home manufacturing companies. These companies routinely exhibit model homes and undertake intensive sales campaigns targeting the general public. If companies that complete 100 to 500 (exclusive) homes annually and those that are

Table 11.8

Number of homes built by home manufacturers in fiscal 1994.

Home manufacturer	Number of units	Share (%)
1 Sekisui Prefab Homes, Ltd.	67,839	4.3
2 Daiwa House Industry	43,592	2.8
3 Misawa Home Co., Ltd.	38,143	2.4
4 Sekisui Chemical Co., Ltd.	30,500	2.0
5 National House Industry Co., Ltd.	26,180	1.7
6 Asahi Chemical Industry Co., Ltd.	14,453	0.9
7 Mitsui Home	10,433	0.7
8 Sumitomo Forestry Co., Ltd.	10,084	0.6
9 Nisseki House Industry Co., Ltd.	9,486	0.6
10 Esu. Bai. Eru.	7,588	0.5
11 Sanwa Home	6,293	0.4
12 Aifuru Home Technology	5,915	0.4
13 Taisei Prefab Construction Co., Ltd.	5,643	0.4
14 Higashi Nippon House	4,571	0.3
15 Taihei House	3,381	0.2
16 Kubota House	3,030	0.2
17 Shokusan Jutaku Sogo Co., Ltd.	2,854	0.2
18 Ichijo Komuten	2,850	0.2
19 Toyota Motor Corporation	2,500	0.2
20 Taisei Corporation	2,401	0.2
Total of 1-10 company performance	258,298	16.6
Total of 1-20 company performance	297,736	19.1
Total of 1-59 company performance[1]	338,852	21.7
Total	1,560,620	100.0

1   Companies that had constructed 500 or more homes.
Source: Nikkei Industrial Information Research Institute (1996), *Nikkei shohin joho;* Ministry of Construction, *Kenchiku tokei nenpou* (fiscal 1995).

considered to be future home manufacturers are included, the number of home manufacturing companies stands at about 70.

In the past, each home manufacturer specialized in its own method of construction. Although an increasing number of major companies now tend to offer all methods, home manufacturers have so far gained only a limited share in the traditional method. In fiscal 1994, home manufacturers accounted for 100% of prefabricated homes and 43% of 2 × 4 homes, but only 6.2% of traditional housing (Ministry of Construction 1995a; Nikkei Sangyo Shouhi Kenkyujo 1996).

Since the 1990s kiln-dried wood and laminated wood have been increasingly used for building materials. This trend accelerated after the Great Hansin-Awaji Earthquake in 1995. Moreover, in response to the Housing

Quality Assurance Law, which was established in 1999, home manufacturers, especially major ones, have increased the use of kiln-dried wood and laminated wood.

The number of homes constructed annually has dropped sharply since 1997 and competition among home builders has become fierce. Small- to medium-sized carpenters and *komuten* have declined in numbers, and even home manufacturers have been experiencing booms and busts.

References
Imported Houses Promotion Association (1994). *Report of the Imported Houses Promotion Association*. Tokyo (in Japanese).
Ministry of Construction (1977). *Housing and Construction in Japan*. Tokyo (in Japanese).
– (1995a). *Questionnaire on Imported Homes*. Tokyo (in Japanese).
– (1995b). *Fiscal 1995 Construction White Paper*. Tokyo (in Japanese).
– (1996). *Fiscal 1996 Annual Report of Construction Statistics*. Tokyo (in Japanese).
– (1999). *Kenchiku tokei nenpou*. Tokyo (in Japanese).
– (2000). *Fiscal 2000 Annual Report of Construction Statistics* (in Japanese).
Nara Prefecture (1996). *Status Quo and Problems of the Lumber Industry in Nara Prefecture* (in Japanese).
Nikkei Sangyo Shouhi Kenkyujo (Nikkei Industry Consumption Research Institute) (1996). *Fiscal 1996 Nikkei Products Information*. Tokyo (in Japanese).
Prime Minister's Office (1996). *Public Opinion Poll on Forests and Forestry*. Tokyo (in Japanese).
Sakai, M., Yukutake, K., and Kojima, M. (1991). *Book of Lumber*. Nihon Ringyo Chosakai, Tokyo (in Japanese)
Yano Keizai Kenkyujo (Yano Economics Research Institute) (1994). Yano Report, May 10. Tokyo (in Japanese).

12
The Japanese Pulp and Paper Industry and Its Wood Use
Hideshi Noda

Since ancient times, paper consumption has been regarded as one barometer of a culture's development. Japan's paper production history can be traced back to the beginning of the seventh century. A papermaking method invented in China in AD 105 spread to Korea in the fourth century and on to Japan in 610. The method was subsequently improved and paper production extended to various places in Japan. Initially, the Japanese traditional handmade paper (*washi*) was so valuable that its use was mostly limited to Buddhist manuscripts. After the 17th century, paper became more widely used for many common goods, such as sliding paper screens (*shoji*), round fans, bamboo-and-paper umbrellas, lanterns, robes, and various craft items. Simultaneously, paper production spread throughout the country, with some regions becoming famous for their paper products. *Washi* is made from bast composed of bark fibres from trees like kohzo (paper mulberry, *Broussonetia kazinoki*) and mitsumata (paperbush, *Edgeworthia chrysantha*).

The process of making *washi* was partially mechanized in the Meiji era (1868-1912). Because the government began issuing monetary notes, the demand for and production of *washi* increased. Japanese production of Western-style paper, begun in the 1870s, also increased rapidly during the Meiji era. Consequently, the supply share of *washi* decreased from about 50% in the 1910s to little more than 10% in the 1930s. The Japanese pulp and paper industry has subsequently modernized by introducing Western papermaking methods. Western-style paper now accounts for the majority of Japanese paper production.

In 1997 Japanese domestic paper production was 31.0 million tons and its consumption was 31.2 million tons, second only to the United States (Figure 12.1). Generally, a country's paper consumption is strongly influenced by its national income level. Japanese per-capita paper and paperboard consumption greatly expanded during the postwar period of rapid economic growth, increasing from 20 kg per person in 1938, the prewar peak, to 245 kg per person in 1997, placing Japan fifth in the world after the US (335 kg), Finland (333 kg), Belgium (301 kg), and Sweden (274 kg).

Figure 12.1

Paper and paperboard production of the top 10 countries in 1997.

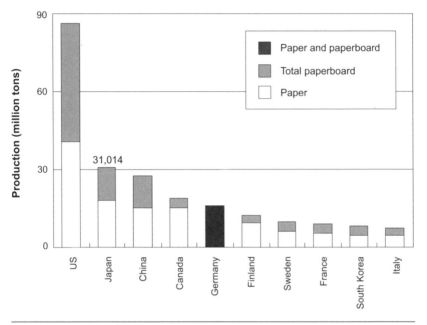

Source: Japan Pulp and Paper Company Ltd. (1998).

Japan's Pulp and Paper Industry: A Brief History

From the Meiji Era to the Second World War

At the beginning of the Meiji era, the Japanese government sent missions to Europe and America to learn advanced Western technologies. Mission members clearly realized that paper manufacturing and the printing industry were necessary to modernize Japan. This led to the importation of Western paper manufacturing technologies and the birth of Japan's modern paper industry in the 1870s. The precursors of Japan's current leading paper companies, Oji Paper Company Ltd. and Nippon Paper Industries Company Ltd., were founded at this time. Newspaper publication boomed and many translated books were published in the 1870s. Because much of the Western-style paper consumed in Japan at this time was imported, and the demand for such paper was increasing, Japanese paper companies began producing it domestically. In 1877, domestic production of Western-style paper was 547 tons, compared with imports of 771 tons. Since Western-style paper was made from domestic raw materials such as rags, straw, and the like, initially it was impossible to produce a large volume of such paper at Japanese mills.

During the 1870s and 1880s, the Japanese paper manufacturing industry was unable to acquire enough raw materials at a sufficiently low cost for its domestically produced Western-style paper to compete with similar imported paper. Japanese manufacturers therefore strove for less expensive production costs by building new mills near forests and their abundant, cheap wood-fibre resources. This led to the adoption of the mass production technology of Western-style paper from wood pulp (sulfite pulp and ground pulp) in 1889-90. Initially, the raw materials were softwoods such as Japanese fir and hemlock growing on the main island of Honshu. In the 1900s, big pulp and paper manufacturers such as Oji Paper and Fuji Paper Company Ltd. constructed new mills on the northern island of Hokkaido, where they started the mass production of paper, particularly newsprint. There, high-quality pulpwood, in the form of yezo spruce (*Picea jezoensis*) and todo fir (Sakhalin fir, *Abies sachalinensis*), was abundantly available. The manufacturers also built mills in Sakhalin, where more of the same pulpwood could be obtained, enabling mass production of paper during the 1910s. Domestic self-sufficiency in Western-style paper was thus established.

As competition between paper companies intensified, many small- and medium-sized companies were driven from the market. Then, in 1933, the three leading manufacturers, Oji, Fuji, and Karafuto, merged into Oji Paper Company Ltd., a huge pulp and paper manufacturer. This company virtually monopolized the Japanese market, producing 80% of the paper and 95% of the pulp. Under such a monopoly and after paper imports peaked at 90,000 tons in 1936, Japanese paper production reached its prewar peak of 1,540,000 tons in 1940.

After the Second World War

The Japanese pulp and paper industry changed greatly after the war. First, because Japan had lost overseas territories like Sakhalin, the industry lost abundant fibre resources and much factory equipment. Paper production decreased to 210,000 tons in 1946. Second, in 1949 Oji Paper was divided into several companies by the US occupation government. Many smaller pulp and paper manufacturers emerged and competition resumed. The prewar monopolistic pulp and paper market was transformed virtually overnight into a competitive industry.

The demand for paper expanded greatly in the rapidly growing postwar Japanese economy. Paper began to be used for various products. Cardboard boxes, for instance, replaced wooden boxes and were widely used as containers for transporting and protecting goods. Thus, paperboard demand increased rapidly. Nationwide, pulp and paper manufacturers expanded production, investing in facilities and equipment and promoting technological innovation (Figure 12.2).

Figure 12.2

Production of paper, paperboard, and pulp, and consumption of wastepaper, in Japan, 1926-97.

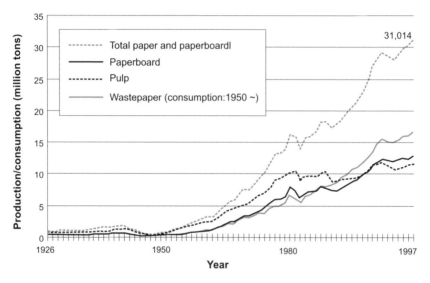

Source: For 1953 onward, Research and Statistics Department, Ministry of International Trade and Industry, *Yearbook of Paper and Pulp Statistics;* for 1926-52, Suzuki (1967).

As previously mentioned, the Japanese pulp and paper industry lost abundant fibre resources after the war. How, then, did the postwar pulp and paper industry procure raw materials? Actually, the history of this industry is characterized by changes in raw materials procurement. Until the first half of the 1950s, the main raw materials for paper were softwoods such as Japanese red pine (*Pinus densiflora*), Japanese black pine (*Pinus thunbergii*), yezo spruce, and todo fir. The demand for pine increased and its price soared because it was also used for construction materials and mine timbers. In the 1960s, therefore, the pulp and paper industry shifted from softwood to hardwood with the adoption of papermaking technologies based on the kraft pulp process (Figure 12.3). The supply of hardwood logs and the combined supply of hardwood logs and chips exceeded the softwood supply in the same categories in 1963.

In addition, the form of supply rapidly changed from logs to chips in the 1960s. In 1955, logs accounted for almost 100% of the pulpwood, but the ratio of chips rose continuously from 24% (1960) to 52% (1965) and to 97% (1997). The importation of raw materials has also increased since the mid-1960s (Figure 12.3). Chips, initially imported from North America by

Figure 12.3

The nature of pulp and paper mills' pulpwood receipts, 1957-97.

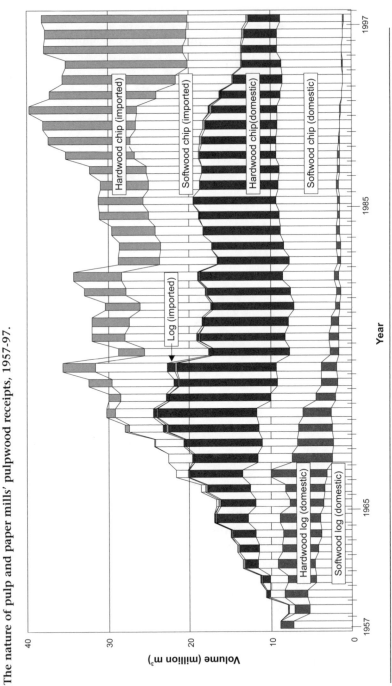

Hardwood chip (imported)

Softwood chip (imported)

Hardwood chip(domestic)

Softwood chip (domestic)

Log (imported)

Hardwood log (domestic)

Softwood log (domestic)

Volume (million m³)

40

30

20

10

0

1957

1965

1985

1997

Year

Source: Japan Paper Association (1999).

chip carriers, are now imported from countries worldwide, including New Zealand, Australia, South Africa, and Chile. Pulpwood imports increased from 19% of the total in 1970 to 38% in 1985 and then 67% in 1997.

We must also consider wastepaper, another papermaking raw material. After the war, wastepaper consumption expanded with its use as a raw material for paperboard, especially containerboard (Figure 12.2). Following the oil shocks of the 1970s and the chip shock of 1980, chip prices rose considerably, causing pulp and paper manufacturers to increase their use of wastepaper to reduce manufacturing costs. Furthermore, advances in de-inking technology in the 1980s increased the amount of wastepaper used for newsprint and wood-containing printing paper (a kind of low-quality printing paper).

As recycling gained popularity, wastepaper consumption reached 16.5 million tons in 1997 (Figure 12.2), and now comprises more than half of the raw material supply for paper. In 1997 the Japanese wastepaper utilization rate was 54% and its recovery rate was 53%, among the highest in the world. Although postwar Japan has been self-sufficient in paper and paperboard production, it has been increasingly less so in raw materials, because the Japanese pulp and paper industry has become dependent on imported raw materials (except wastepaper, which is still from domestic sources). Thus, the relationship between the Japanese pulp and paper industry and Japanese forestry has become both weaker and less direct.

Demand for Paper and Paperboard

Main Uses for Paper and Paperboard

The final uses for paper and paperboard in Japan are graphic use (printing and communication), packaging and industrial use (bags and paperboard boxes), and sanitary use. Table 12.1 shows the changes in consumption for each use. Consumption of paperboard, particularly containerboard, increased during the period of high economic growth (the 1960s through the first half of the 1970s). Although packaging and industrial use once accounted for 60% of the total consumption, that share has constantly decreased. In contrast, consumption for graphic use has steadily expanded, reaching 47% in 1997.

Paper and paperboard import and export quantities are extremely small in Japan's nearly self-sufficient market (Table 12.l). One reason for the small import amount is that Japanese consumers are choosy about their paper and paperboard quality, and in some product categories only domestic paper manufacturers have been able to produce paper of sufficiently high quality. Another reason is that the Japanese paper and paperboard distribution system remains inefficient and partially closed, in spite of the import tariff reduction on paper and paperboard. (See "Supply of Paper and Paperboard" below.)

Table 12.1

Consumption, exports, and imports of paper and paperboard in Japan, 1965-97.

Consumption[2]	1965 Volume[1]	%	1975 Volume[1]	%	1985 Volume[1]	%	1997 Volume[1]	%
Graphic use								
Newsprint	1,156	17	2,036	15	2,693	13	3,598	11
Printing and writing	1,273	18	2,736	21	4,643	23	9,476	30
Communication, etc.	–	–	689	5	1,238	6	1,824	6
Total	2,429	35	5,461	41	8,574	42	14,898	47
Packaging and industrial use								
Packaging and wrapping	833	12	1,043	8	1,107	5	1,103	4
Miscellaneous	542	8	440	3	1,010	5	1,160	4
Paperboard	2,851	41	5,756	43	8,421	42	12,585	40
(Containerboard)	–	–	(3,962)	(30)	(5,807)	(29)	(9,406)	(30)
(Boxboard)	–	–	(1,200)	(9)	(1,613)	(8)	(2,102)	(7)
Total	4,226	61	7,239	54	10,538	52	14,848	47
Sanitary use	329	5	639	5	1,089	5	1,706	5
Total consumption	6,984	100	13,339	100	20,201	100	31,452	100
Exports (paper and paperboard)	230	3	654	5	920	5	964	3
Imports (paper and paperboard)	17	0	107	1	705	3	1,323	4

1 In thousands of metric tons.
2 Consumption = Shipment + Imports – Exports.
Source: Japan Paper Association, *Pulp & Paper*, no. 564, 1996, p. 8; for 1997, Japan Paper Association (1998a).

Newsprint

Although newsprint consumption increased in the growing Japanese economy, its percentage of total paper consumption gradually decreased to 11% in 1997 (Table 12.1). To save fibre resources and to lower printing and transport costs, newsprint weight has been reduced since the 1970s oil shocks. In 1997, 88% of newsprint was super-lightweight paper (43 g/m^2), making it 17% lighter than the standard weight (52 g/m^2) in 1975. Besides being lightweight, newsprint is expected by newspaper publishers be strong enough for high-speed printing and colour printability. In addition, different publishers demand different newsprint qualities. As previously noted, Japanese consumers are very choosy about paper and paperboard quality. To satisfy all consumers' different needs, paper manufacturing companies must therefore produce paper of different but high qualities.

Only a small amount of newsprint is imported from North America and Northern Europe. Japanese paper manufacturers have, however, recently begun to produce high-quality newsprint overseas through joint ventures with foreign companies. Imports of such newsprint increased gradually to 628,532 tons in 1997, half of total paper and paperboard imports. Newsprint imports became duty-free in 1990.

Printing paper

The printing industry consumes large quantities of printing paper. Printing and writing paper are widely used in office communications, books, magazines, notebooks, catalogues, handouts, and equipment manuals. The demand for printing paper has increased, with current consumption accounting for 30% of total paper and paperboard consumption.

Printing paper is classified as either coated or uncoated paper. Consumption of coated paper with colour printability has been increasing in recent years, and represents the majority of printing paper today. In particular, the demand for lightweight coated paper has grown substantially, with production expanding from 479,530 tons in 1988 to 1,334,372 tons in 1997. In 1996 there were about 17,200 printing companies in Japan, most of them small- or medium-sized enterprises.

Containerboard

Containerboard is made into corrugated board, which is used in packaging a wide variety of goods, over half of them food products. Most notably the demand for corrugated board for packaging processed foods has been increasing, accounting for 35% of total corrugated board demand in 1997. The second largest demand, for packaging fruits and vegetables, accounts for 15% of total demand. Corrugated board is also used for packing electrical and mechanical apparatus (12%), pottery and miscellaneous goods (7%), and medicine and cosmetics (6%).

Table 12.2

Production of different grades of paper and paperboard in Japan, 1965-97.

Grade	1965 Volume[1]	1965 %	1975 Volume[1]	1975 %	1988 Volume[1]	1988 %	1997 Volume[1]	1997 %
Paper[2]								
Newsprint	1,184	16	2,160	16	3,067	12	3,192	10
Printing and Communication	1,211	17	2,772	20	–	–	–	–
Printing and Writing	–	–	–	–	6,440	26	9,251	30
(Uncoated)	–	–	–	–	(3,161)	(13)	(3,138)	(10)
(Lightweight Coated)	–	–	–	–	(480)	(2)	(1,334)	(4)
(Coated)	–	–	–	–	(2,468)	(10)	(4,388)	(14)
Communication	–	–	–	–	1,160	5	1,841	6
Packaging and Wrapping	806	11	1,037	8	1,129	5	1,108	4
Household Tissue Paper	450	6	622	5	–	–	–	–
Sanitary Tissue	–	–	–	–	1,281	5	1,715	6
Miscellaneous	–	–	909	7	1,266	5	1,160	4
Other[3]	568	8	211	2	–	–	–	–
Total	4,219	58	7,711	57	14,343	58	18,268	59
Paperboard								
Containerboard	1,795	25	4,037	30	7,103	29	9,425	30
(Linerboard)	(1,126)	(15)	(2,617)	(19)	(4,416)	(18)	(5,733)	(18)
(Corrugating Medium)	(669)	(9)	(1,419)	(10)	(2,687)	(11)	(3,692)	(12)
Boxboard	900	12	1,240	9	1,958	8	2,236	7
(Whiteboard)	(597)	(8)	(952)	(7)	(1,602)	(7)	(1,955)	(6)
Miscellaneous	384	5	613	5	1,220	5	1,086	4
Total	3,079	42	5,890	43	10,281	42	12,747	41
Total paper & paperboard	7,299	100	13,601	100	24,624	100	31,014	100

1 In thousands of metric tons.
2 Paper classification has been altered since 1988 (dashes indicate that a particular grade was not applicable during that year).
3 "Other" included Japanese paper (*washi*) in 1965.

Source: Research and Statistics Department, Ministry of International Trade and Industry (various years), *Yearbook of Paper and Pulp Statistics.*

Corrugated board, an excellent packing material, is light, easy to print, and recyclable. Gaining wide use, it rapidly replaced wooden packing materials after the Second World War. Since the collapse of the "bubble economy," however, demand has remained sluggish due to the decrease in distribution and the trend towards reduced packaging. To create new demand, paperboard manufacturers have developed new types of paper corresponding to customers' new needs, such as recycled pulp mould made from wastepaper. Although pulp mould is different from corrugated board, it is increasingly replacing Styrofoam and is used for packing home electrical products. Many of the 3,100 corrugated board manufacturers (1996) are also small- or medium-sized enterprises.

Supply of Paper and Paperboard

Overview of the Production and Distribution System

Table 12.2 reflects the increase in the production of printing papers (especially coated paper) and containerboard between 1965 and 1997. Japanese import tariffs on paper and paperboard were incrementally reduced because of increased foreign demand for improved access to the Japanese paper market (Table 12.3). Thus, the average tariff rate on paper and paperboard was lowered to 1.2% in 1997. Meanwhile, with a strong yen since 1985, the Japanese pulp and paper industry has actively promoted investment abroad. It is therefore assumed that future paper imports, especially of newsprint, will gradually increase.

Table 12.3

Japanese import tariffs on paper and paperboard, 1980-99.

Main grade	Import tariff (%)					
	1980	1985	1990	1995	1997	1999
Newsprint	4.3	3.9	0.0	0.0	0.0	0.0
Woodfree [1]	6.0	5.8	4.6	4.6	4.1	2.9
Coated [1]	7.8	7.2	4.1	3.7	2.9	2.1
Mechanical coated [2]	7.3	5.1	4.1	3.7	2.9	2.1
Linerboard [3]	12.0	6.5	3.5	3.2	2.1	1.8
Corrugating Medium	12.0	12.0	9.6	9.6	8.4	6.0
Whiteboard [4]	8.0	5.0	2.5	2.3	1.5	1.3
Milk Carton	7.3	4.8	0.0	0.0	0.0	0.0
Average [5]	7.3	5.3	1.9	1.7	1.2	

1 150 grams per m^2 or less.
2 Including lightweight coated paper (LWC).
3 300 grams per m^2 or less.
4 More than 300 grams per m^2
5 Based on import value.
Source: Japan Paper Association (1998a, 1998b); for 1999, Japan Paper Association, *Pulp & Paper*, no. 603, p. 12, 1999.

Figure 12.4

Outline of paper and paperboard supply chain and main users in Japan.

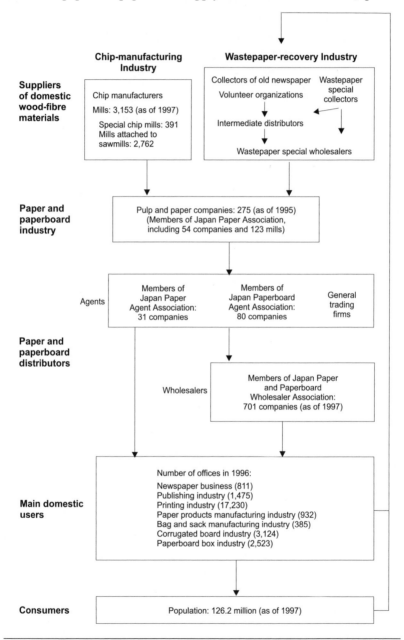

Source: Japan Paper Association (1998a); Japan Pulp and Paper Company Ltd. (1998); Statistics and Information Department, Ministry of Agriculture, Forestry, and Fisheries, 1997, *Report on Demand and Supply of Timber.*

Figure 12.4 illustrates Japan's paper and paperboard production, distribution, and consumption channels. The chip-manufacturing industry and wastepaper-recovery companies are the main suppliers of domestic raw materials. There were 275 pulp and paper manufacturing companies in 1995; the major companies are members of the Japan Paper Association, which dates back to 1880 and consists of 54 companies and 123 mills. The 25 largest companies in Japan accounted for 82% of the total paper and paperboard production in 1997.

There are two types of distributors: (1) agents who buy products directly from paper manufacturing companies and sell them to wholesalers or consumers, and (2) wholesalers doing business between those agents and consumers. Paper and paperboard are distributed to consumers through both agents and wholesalers. Agents (about 90 companies) handle the products of specific paper manufacturing companies. They have nationwide sales networks and provide paper and paperboard to wholesalers and directly to big consumers (e.g., national newspaper publishers, big publishing companies, and large printing firms). There are about 700 wholesalers with local sales networks. They mainly buy paper and paperboard from agents and sell to small- and medium-sized consumers in the regional printing, publishing, and paperboard-box industry.

The Japanese paper and paperboard distribution system is inefficient and partially closed because of old business practices. Efficiency could be improved by streamlining the distribution system, improving information networks, and reorganizing paper manufacturers and distributors through a process of merger and acquisition (M & A).

Major Japanese Paper Manufacturers

The production systems of Japan's dominant paper manufacturers are integrated from pulp through to paper and paperboard production. In 1995 the producing companies' integrated mills consumed 86% of total pulp production. Some paperboard manufacturers who use wastepaper, however, produce very little pulp at their mills. No Japanese firm specializes in pulp production only. As previously described, intense competition among paper manufacturers resumed after the war, leading to a vicious industry cycle of "demand boom – production increase – capital investment – excessive production – recession – market stagnation – profit deterioration – curtailment of operations and abandonment of excessive equipment." The paper manufacturers' financial condition has greatly deteriorated in recent years due to a downturn in this cycle caused by the recession after the collapse of the "bubble economy." Paper manufacturers are once again trying to strengthen their management through merger and acquisition.

Japanese paper manufacturers began reorganizing through two successive large-scale mergers in 1993 that created two huge companies, the Nippon

Paper Industries Company Ltd. and the New Oji Paper Company Ltd. In 1995 New Oji Paper produced 3.6 million tons of paper and paperboard (12.0% production share) and Nippon Paper 3.0 million tons (10.1% share). Subsequently, Oji Paper Company Ltd. was formed when New Oji Paper and Honshu Paper Company Ltd. merged in 1996. In 1997 Oji Paper's paper and paperboard production was 6.0 million tons (19.4% production share) and sales were about ¥980 billion. Currently, Oji Paper is the largest company in the Japanese pulp and paper industry, ranking second in worldwide sales in 1997.

Japanese paper and paperboard market shares held by major companies became increasingly concentrated after these 1990s mega-mergers. In 1985 the production share of the three largest paper and paperboard companies was 25% and that of the 10 largest companies was 50%. By 1997 these had risen to 38% and 63%, respectively. Concentration rates are even higher in the newsprint market. In 1997 the production share of the three largest companies in the following sectors was 68% in the newsprint market, 64% in the business communication paper market, 59% in the coated paper market, and 54% in the packaging paper market. All of these contrast sharply with the low concentration (31% in 1997) in the paperboard market.

These Japanese mega-mergers coincided with the worldwide pulp and paper industry M & A trend of the 1990s. To survive in the age of global competition, when deregulation and trade liberalization are promoted worldwide, European and American pulp and paper industries are also active in M & A, aimed at reinforcing competitiveness. Thus, Japanese paper manufacturers must compete fiercely with powerful overseas manufacturers in both the domestic market and the expanding Asian markets. To enhance its competitiveness and survive in the world market, the Japanese pulp and paper industry must reorganize itself.

Current Supply of Raw Materials for Paper Manufacturing

Pulpwood

In 1997 the total pulpwood supply was 38.1 million m³, with chips accounting for 97% of the total (Figure 12.3). To secure a stable supply of raw materials, the Japanese pulp and paper industry recently began importing chips from many countries. Sixty-eight percent of chips were imported in 1997, with 33% of the total being US hardwood chips (oak and other species), 25% from Australia (eucalyptus), 12% from Chile (beech and eucalyptus), and 11% from South Africa (eucalyptus and *Acacia mangium*). Forty-four percent of softwood chip imports were from the US (Douglas-fir wood residue and spruce from mills), 28% from Australia (Monterey pine and other species), and 8% from Canada (spruce, pine, and fir wood residue from mills). Generally, imported chips are supplied to the pulp and paper manufacturers through trading firms and are transported by 40,000-ton class chip carrier ships. By the end of 1997, there were 88 of these ships.

To secure a longer-term raw material supply, and in cooperation with trading companies in South America, Oceania, and Southeast Asia, Japanese paper manufacturers promoted large-scale eucalyptus and *Acacia mangium* afforestation in the 1990s. From the environmental protection perspective, it is extremely important that paper manufacturers regenerate wood fibre resources by planting trees. Although some manufacturers have company-owned forests in Japan, a comparison of 15 companies put the average forest size at only about 18,000 ha. One reason is that both afforestation and reforestation costs are extremely high in Japan.

Still an important supply source although its share is decreasing, domestic pulpwood accounted for only 33% (12.7 million m^3) of the total supply in 1997. Pulpwood is processed into chips at chip mills (3,173 mills in 1997) and supplied to paper manufacturers, as shown in Figure 12.4. Many chippers are attached to the sawmills (2,762 such sawmills in 1997), which generated 62% of domestic chips. Most chips are processed from softwood residue. Alternatively, special chip mills (391 such mills in 1997) chiefly process hardwood logs into chips. In recent years, the profitability of Japanese chip companies has declined rapidly due to the increase of imported chips and sluggish chip prices. Chip mill numbers decreased drastically from 4,494 in 1990 to 3,173 in 1997. Special hardwood chip mills in particular have suffered because domestic hardwood chips have lost price competitiveness with imported chips due to the higher yen value in the 1990s. The Japanese chip industry must restructure itself to become more competitive.

Wastepaper

Wastepaper recycling increased rapidly after the Second World War and wastepaper now accounts for much of today's raw material for papermaking. In 1997 wastepaper consumption was 16.5 million tons. Wastepaper is classified into nine grades in the Paper and Pulp Statistics Yearbook (Research and Statistics Department 1998). Regarding the major grades, old corrugated board accounted for 46% of total wastepaper consumption, old newspaper 23%, and old magazines 14%.

Wastepaper comes from many sources, including homes (newspapers and magazines, etc.), supermarkets and shops (corrugated fibreboard containers), offices (business communication papers, etc.), and printing offices and paperboard-box mills (industrial wastepaper). The wastepaper recovery system is therefore quite complex (Figure 12.4). Wastepaper collectors recover old newspapers and magazines from homes in exchange for toilet paper. Volunteer organizations such as town associations also recover wastepaper. Wastepaper wholesalers collect old newspapers and magazines through intermediate distributors and deliver them to the paper and paperboard manufacturers after separating and packing them by grades. Wastepaper collectors (including some wholesalers) recover considerable wastepaper from printing offices, shopping centres, office buildings, and so on, delivering it to wastepaper wholesalers.

In 1998 the average wastepaper recovery rate in Japan was 55%. The recovery rates by grades were: 113% for old newspapers (it exceeds 100% because inserted advertisements are included in old newspapers), 78% for corrugated board, and 31% for old printing and communication papers. Theoretically, the average recovery rate could be improved to 66%. Thus, in the future the Japanese wastepaper recovery industry faces the challenge of improving the wastepaper recovery rate for old printing and communication papers.

Tasks of the Japanese Pulp and Paper Industry

To survive in the age of global competition, the Japanese pulp and paper industry must further reorganize and streamline the paper and paperboard distribution system. Moreover, the industry should improve both utilization and wastepaper recovery rates, and expand tree plantations to regenerate wood-fibre resources. These efforts are vital for environmental protection.

In that context, the Japan Paper Association launched the "Self-Imposed Action Plan for the Environment" in January 1997, whereby the industry committed itself to the following goals: (1) reduce energy consumption per ton of paper produced by the year 2010 to 90% of the 1990 level, and (2) expand foreign and domestic tree plantations to 550,000 ha by 2010. Targets were also set to establish a recycling-based society: (1) to boost the wastepaper utilization rate to 56% by 2000 (a 57% rate was achieved) and (2) by the year 2010, to reduce industrial wastes going to landfills, per ton of paper produced, to 40% of the 1990 level. The program included measures to cope with environmental risks from small amounts of toxic chemicals (such as dioxin, formaldehyde, and benzene) and to establish an environment management system (Japan Paper Association 1998b).

Finally, it is important for economic survival and responsible environmental stewardship that the Japanese pulp and paper industry transform itself from an extractive industry, based on mass consumption of natural resources, into a sustainable industry, based on recycling and regeneration of fibre resources.

References
Japan Paper Association, ed. (1998a). "Kami-parupu sangyo no genjoh," *Kami-Parupu (Pulp & Paper)*, no. 590. 31 pp. (in Japanese).
– (1998b). *Pulp & Paper Statistics 1998*. 25 pp. Japan Paper Association, Tokyo.
– (1999). *1999 Parupu-Zai Binran*. 110 pp. Japan Paper Association, Tokyo (in Japanese).

Japan Pulp and Paper Company Ltd., ed. (1998). *Zuhyo: Kami-Parupu Tokei 1998*. 50 pp. Japan Pulp and Paper Company Ltd., Tokyo (in Japanese).

Oji Paper Company Ltd., ed. (1993). *Kami-Parupu no Jissai-Chishiki*. 5th ed. 209 pp. Toyo Keizai, Tokyo (in Japanese).

Paper, Pulp and Printing Industry Division, Ministry of International Trade and Industry, ed. (1994). *Ryokka to Kokusaika no-nakano Kami-Parupu Sangyo*. 310 pp. International Trade and Industry Research Center, Tokyo (in Japanese).

Research and Statistics Department, Minister's Secretariat, Ministry of International Trade and Industry, ed. (1998). *1997 Yearbook of Paper and Pulp Statistics*. 212 pp. International Trade and Industry Statistics Association, Tokyo (in Japanese).

Saiseishi Kaihatsu Team, Honshu Paper Company Ltd., ed. (1991). *Kami no Risaikuru Hyaku no Chishiki*. 219 pp. Tokyo Shoseki Company Ltd., Tokyo (in Japanese).

Suzuki, Hisao, ed. (1967). *Gendai-Nippon-Sangyo Hattatsushi XII Kami-Parupu*, p. 14 (in Appendix Table of Statistics). Gendai-Nippon-Sangyo Hattatsushi Kenkyu-Kai, Tokyo (in Japanese).

13
Local Forestry and Sawmill Industries: The Case of Kumano, Mie Prefecture
Kozue Taguchi

This chapter presents a case study of the forestry and sawmill industries in a specific area, so that forestry and sawmill industries at a regional level may be understood from actual examples. The area is that of Kumano in Mie Prefecture. Based on field and desk research, the structure of the forestry and sawmill industries in Kumano is examined and their characteristics and underlying factors are discussed.

The forestry and sawmill industries in different regions show different characteristics, reflecting local natural conditions, local history, and social or economic conditions in the region. At the same time, local operations are influenced by the general trends in the forestry and sawmill industries in Japan. This chapter describes the forestry and sawmill industries in Kumano not only in isolation but also from the wider perspective of the history and current situation of the forestry and sawmill industries in Japan. It is intended that these industries as a whole will be better understood through an understanding of the situation in Kumano.

Kumano is situated in the southeastern part of the Kii Peninsula, facing the Pacific Ocean (Figure 13.1). Although the name Kumano can refer to a larger area along the Kumano River that extends across the borders of three prefectures, it is used here to refer to Kumano City and the neighbouring towns of Minamimuro County in Mie Prefecture. Kumano has some of the highest rainfall in mainland Japan, and the climate is quite mild throughout the year. Kumano is located on the borders of three prefectures: Mie, Nara, and Wakayama, all of which have traditionally been active in forestry. Forestry is important to the economy of Kumano, representing 6% of the region's total industry; this is a much higher figure than the average of 0.7% for the prefecture as a whole (Owase Kumano Ryuiki Ringyo Kasseika Center 1997b, 5-6).

The first section briefly discusses the history of the forestry and sawmill industries in Japan, and their related history in the Kumano region. The next section details the characteristics of the forestry and sawmill industries

Figure 13.1

Location of Kumano.

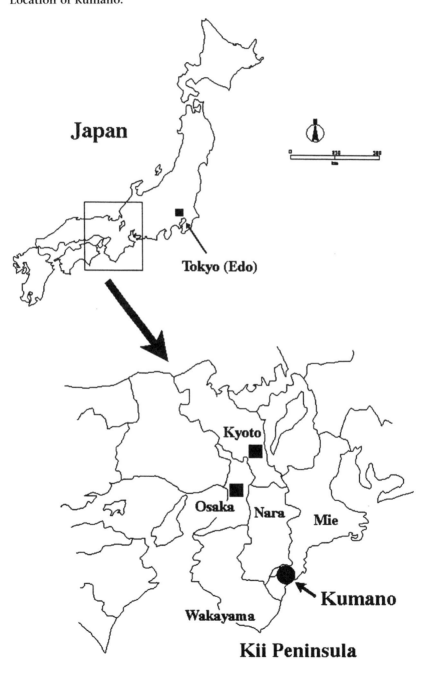

in Kumano today. The final section discusses the factors supporting these industries in Kumano, and focuses on the relationship between the two industries.

Historical Background

Before the Second World War

Since the humid and warm climate of much of Japan is suitable for silviculture, the planting of sugi (Japanese cedar, *Cryptomeria japonica*) and hinoki (Japanese cypress, *Chamaecyparis obtusa*) began as early as the 16th century, and forestry developed in many areas. The traditional forestry areas developed through trade with big consumer cities, such as Osaka, Kyoto, and Edo (the former name of Tokyo). To meet the demand from these cities, they produced specialty timber of high quality, using traditional silvicultural methods suited to the local natural conditions. These traditional forestry areas were well known as timber-producing areas for a particular local high-quality specialty. The most famous timber-producing area was Yoshino in Nara Prefecture. Here, a labour-intensive silviculture of long rotation was developed, in which 10,000 sugi and hinoki seedlings were planted per hectare, and frequent pruning and thinning operations were performed to produce high-quality timber.

Kumano was also part of a timber-producing area called Kishu in the Edo period (1603-1867). This region facing the Pacific Ocean offered transportation advantages, since wood was transported mainly by sea. Large quantities of wood were delivered from Kishu to Edo, the capital and a big consumer. It has been estimated that planting began in Kumano in the 1670s (Mie-ken 1988, 868), and Kumano can be considered one of the traditional forestry areas. The silviculture in Kumano's forestry was very extensive, however, in contrast to that of Yoshino.

Kumano's rich, nutritious soil and warm, wet climate make it very suitable for growing trees, especially sugi. Trees in Kumano's forests grow better and faster than in other forestry areas on the main island, and the forests can produce large trees and therefore lumber of large dimensions. The traditional silviculture that developed in Kumano's natural conditions was the cultivation of low-density sugi plantations. It was represented by a method of forestry called *nasubigiri* (eggplant harvesting), which was performed until the war in the mountain area along the Kitayama River. It was a selective method in which the best-grown sugi trees were harvested first (just as in the harvesting of eggplant), followed by replanting of the harvested sites. *Nasubigiri* was characterized by a low planting density at 2,000 seedlings per hectare, a long rotation of 60 to more than 120 years, and the harvesting of large mature trees. There was almost no pruning or tending, and only a little weeding. Harvested logs were floated down the river to the harbour on the Pacific Ocean, and then transported mainly to Edo and Osaka. It is said that this selection method was developed because only large trees could justify the transportation costs associated with being floated down the river (Mie-ken 1988, 872). The large sugi trees from Kumano were processed and used for housing, ship construction, and barrels for the Japanese rice wine, *sake*. In those days, a wide variety of wooden products other than construction timber were produced from timber of local specialties in these traditional forestry areas.

After the Second World War
Forestry areas in Japan have developed quite differently since the war, in the wake of drastic changes in society and the economy. Large quantities of cheap imported wood have had a big impact on forestry and the forest products industry. Since 1961, when timber imports were first liberalized, the share of domestic wood in the market has shrunk. In particular, second-grade lumber of sugi, one of the two major plantation species in Japan, came into direct competition with cheaper imported wood such as hemlock (*Tsuga heterophylla*) and lost much of its market. Many timber-producing areas began specializing in the production of first-grade lumber called *keshozai* or *yakumono*, from high-quality logs, and to develop themselves as *sanchi*. (*Sanchi* was originally a term in agriculture, and refers to an area known for producing certain agricultural products.) *Sanchi* have typically been seen more as lumber-producing areas (Handa 1986, 23), whereas before the war traditional forestry areas were known as timber-producing areas.

Typical high-quality logs are straight and knot-free, with narrow and even annual rings. Such logs make beautiful, first-grade lumber, and are therefore highly appreciated in traditional Japanese housing, in which much of wood is left exposed; they have an interior-decorating as well as a structural function. In other words, the strategy of many *sanchi* areas was to compete with imported wood on quality rather than price. This was possible because there was a demand for first-grade lumber from consumers, whose living standards increased with Japan's economic growth, and the price of first-grade lumber remained relatively high.

Many *sanchi* sawmills are specialized, and produce the specialty of the particular *sanchi*. The most typical product is posts of hinoki, as this wood has been favoured as being both strong and beautiful. This reputation has resulted in markedly higher prices being paid for hinoki than for sugi. In addition, there had been almost no competing imported wood species able to replace hinoki in the market, because of its unique appearance and strength. Hinoki has enjoyed demand since the late 1960s, especially in big cities, and this has pushed up its price. Many timber-producing areas have therefore tried to specialize in the production of hinoki posts of first-grade lumber (Handa 1986, 18).

Sanchi are still primarily based on local good-quality forest resources, from which they produce a certain specialty product. Before the war, the sawmill industry used mostly local logs suited to their specialty products, but the development of log auction markets after the war and the resulting expansion of log distribution has led to *sanchi* using logs from a wider area. Sawn products from *sanchi* are also widely distributed, some nationwide. *Sanchi* today are therefore characterized by their concentration of specialized sawmills producing a local specialty, with widespread sourcing of raw materials and extensive distribution of their products. However, the size of *sanchi* varies; for example, Tono, a famous *sanchi* producing hinoki posts, consists of about 200 sawmills, consuming 200,000 m^3 of logs in 1996, two-thirds of which was domestic wood (Handa 1999, 161).

Not many areas have been able to develop themselves into a *sanchi*, however, and only a limited number of forestry areas have been able to provide logs to *sanchi* (Morita 1994, 191). Many forestry areas declined as their specialty products were replaced by imported wood or substitutes. In some areas, both the forestry and sawmill industries declined, while in other areas the local sawmill industry began using imported wood as their raw material and the local forestry industry declined (Chiiki Ringyo Kenkyukai 1978, 90).

Although it was a traditional forestry area, Kumano never developed as a *sanchi* after the war. Before the war, logs produced in Kumano were floated down the river to another area, so a sawmill industry did not develop in Kumano. After the war, logs could no longer be floated downstream because of dam construction, but forest road construction made land transportation of forest products possible and a sawmill industry developed in Kumano. The selection method, *nasubigiri*, quickly declined with the end of log floating and clear-cutting became prevalent, with planting densities being increased.

Since prices for hinoki have risen, hinoki has often been chosen for planting after the harvesting of sugi. Most of the timber from Kumano's forests was still sugi, however, grown to a large size under an extensive silvicultural system. Such timber generally has wide annual rings and is therefore unsuitable for the production of first-grade lumber unless it is old and has a large diameter. Much of the abundant old large timber, grown under *nasubigiri* before the war, was harvested during the war for ship construction, so much of the timber available in Kumano's forests after the war was relatively young and suitable only for second-grade lumber.

Kumano's forestry and domestic sawmills have not totally declined, however. Domestic sawmills have managed to maintain their business, although on a rather small scale, and forestry has been practised continuously. The next section examines the structure of the forestry and sawmill industries in Kumano.

The Forestry and Sawmill Industries Today

Forestry

Forest Resources and Ownership
At the centre of forestry in this mountain area is the Kumano City district, which has about 23,000 ha of forests covering 87% of the area. This high rate of forest coverage reflects the fact that most of the area is mountainous, except for the seashore. About 85% of forests are privately owned, and 85% of these private forests were planted. In the mountains, trees were artificially planted wherever planting was possible. These man-made forests consist of sugi and hinoki, with sugi having a little more growing stock at 55% by area (Kumano-shi 1995, 2). Figure 13.2 shows that in the Kumano forests, there are less than average very young trees and more old trees compared with the national average. Small-scale ownership is dominant among the total of 1,402 forest owners, as is the case in the entire country, with 80% owning less than 5 ha of forests and only 1% owning more than 50 ha of forests (Kumano-shi 1995, 4).

Silviculture
The average planting density today is 4,000-5,000 seedlings per hectare,

Figure 13.2

Age-class distribution of man-made forests in Japan and in Kumano City, 1990.

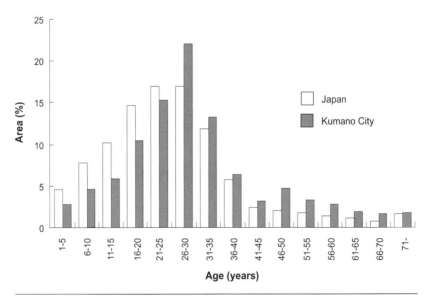

with a rotation period of about 50-70 years for sugi and 60-100 years for hinoki. After the late 1960s, when first-grade lumber started to become popular, the focus of silviculture changed to the production of high-quality logs: planting density increased, hinoki replaced sugi as the plantation species of choice, and pruning, which had not been done previously in Kumano, began. There has been less pruning recently, however, due to the ever-declining profitability of the forestry business and the decreasing demand for high-quality hinoki. Instead, for reasons of silvicultural cost saving, the production of large sugi trees is now being reconsidered. As mentioned previously, Kumano's mild climate, frequent rainfall throughout the year, and very rich soil together help sugi grow faster and better than in other areas. The most recent idea, yet to be put into action, is to take advantage of the natural conditions and again grow large sugi trees economically.

Planting and tending practices have decreased in Kumano as they have elsewhere due to the declining profitability of forestry. In fiscal year 1995, however, 22% of the area planted in private forests in Mie Prefecture was in the Kumano area, which has 12% of the prefecture's total forest reserves (Mie-ken 1997, 44). A simple comparison of these two figures suggests that planting is more active in Kumano than in Mie Prefecture as a whole. The general tendency is for harvested forestland, especially in remote mountains, to be abandoned without new trees being planted, but in Kumano harvested forestland is usually replanted. The profitability of thinning is very low, even in forestry areas that produce high-quality logs. Tending and thinning have both been neglected in Kumano's forests, as in many other areas. If there is to be a shift to the silviculture of large sugi, however, as is being reconsidered, then a way needs to be found to promote thinning by making it profitable.

Forestry Labour and the Forest Owners' Association
In the Kumano area there are about 250 forestry workers, but this number decreases by about 30 every year and the average age is increasing. According to a survey conducted in the Kumano area by Mie Prefecture, the average age of a forestry worker is 56, and 74% of workers are over 50 years old, compared with the national average of 69% (Owase Kumano Ryuiki Ringyo Kasseika Center 1997a, 30; Forestry Agency 1996, 68).

The Kumano forest owners' association has tried to secure more young workers by improving working conditions and providing training, and has tried to promote thinning in members' forests. The association introduced a monthly guaranteed salary and welfare system, and provides a training program for newcomers. To promote thinning, the association built a factory in 1997 to process thinned wood into products such as lamina for glue-laminated wood.

Log Production and Log Auction Markets

In 1995 log production in the Kumano area was 46,000 m³, almost all of which was harvested by clear-cutting (Owase Kumano Ryuiki Ringyo Kasseika Center 1997b, 11, 27). Logs are produced by individual log production enterprises and by the forest owners' association, with the latter having a 10% market share (Owase Kumano Ryuiki Ringyo Kasseika Center 1997a, 40).

Since its foundation in 1982, the Kumano Log Auction Market, the only one in Kumano, has played an important role in integrating local log production and sawmills. The log auction market is located in the mountain area where forestry is conducted and sawmills are situated. An auction is held twice a month, and 25,000-30,000 m³ of logs are handled each year, providing the market with a profit of ¥1-1.3 billion annually. About 70% of the logs handled at the market are sugi, and 30% are hinoki.

Most of the logs brought into the market are from local forests (60% by volume). Although the logs are more diversely distributed through the market, 40% are still bought and processed by local sawmills. The volume of logs handled at the market represents about two-thirds of the total log production in the area (46,000 m³ in 1995). The rest are either sold directly to sawmills by log producers or transported out of the Kumano area. As mentioned previously, the Kumano district is located on the borders of three prefectures, and many log-producing enterprises from neighbouring prefectures come to Kumano forests and then take the harvested logs back to their own areas. Sawmills in neighbouring prefectures also come to the Kumano Log Auction Market to procure logs.

Logs vary in diameter, size, and age, but logs from large and valuable trees that are more than a hundred years old are sometimes brought into the market. They are considered the fruit of a long tradition of forestry in Kumano, where selection methods based on long rotation were used until the war. Generally speaking, the market has a good reputation with the local sawmills for the quality of the logs sold there, and this suggests that the market attracts some of the better-quality logs produced in the area. An anticipated problem is that the logs brought into the market will be from younger trees in the future, since the average age of local forests is decreasing.

Following the establishment of Japan's first log auction market in the 1950s, the number of newly opened markets increased until 1980, after which it declined (Morita 1994, 120). The log auction market in Kumano was founded later than most other log markets in Japan. This was probably because the sawmill industry developed late in Kumano, and log auction markets had already been opened in neighbouring *sanchi* areas. Until the auction market was opened in Kumano, much of the timber produced there was sent to auction markets, dealers, and sawmills in neighbouring areas. The log producers and sawmills in Kumano that previously used these markets to sell and buy logs also traded logs directly with each other. The Kumano Log Auction Market, therefore, now functions as a collection point for local logs that might previously have been exported from the district, but which are now sold primarily to local sawmills.

Sawmill Industry

As of 1995 there were 34 sawmills in the Kumano area, of which 26 processed domestic wood and 8 used imported timber. None of these sawmills handled both domestic and imported materials. Total wood consumption was about 110,000 m³, of which 70% was imported and 30% was of domestic origin. Many of the sawmills processing imported wood used to produce different kinds of board from medium-sized sugi logs. They switched as boards were replaced by plywood made from imported wood.

The domestic sawmill industry in Kumano consists of 26 sawmills consuming 30,000 m³ of logs, a smaller volume than for *sanchi*. The average log consumption per sawmill is 1,150 m³, again smaller than the national average of 1,867 m³ (Ministry of Agriculture, Forestry, and Fisheries 1996, 39). According to a survey by Mie Prefecture, 70% of the domestic logs used are from the local forests in Kumano and Owase, an area to the north of Kumano, and 20% are from neighbouring areas in Nara and Wakayama prefectures (Owase Kumano Ryuiki Ringyo Kasseika Center 1997a, 19).

The log auction market plays the most important role in supplying logs to the sawmills. For example, the six sawmill owners in Kumano City interviewed by the author procure their logs mainly through the log auction market in Kumano. Most of the sawmills using domestic materials produce lumber for housing construction. Among them, about 10 mills do business only locally with carpenters and housing companies, and the others sell their products to "lumber markets" near cities. The former are retail-based, whereas the latter are based on the wholesale market for lumber. "Retail" here basically means that the products are sold to local carpenters and small housing companies for building houses, rather than being sold directly to end-consumers, as in do-it-yourself stores.

One of the characteristics of the Kumano sawmills processing domestic timber is their high percentage of retail sales. Compared with a neighbouring area, Owase, which is a *sanchi* famous for its first-grade hinoki posts, the sawn products of Kumano are sold more locally, whereas the majority of Owase's products are sold to outside lumber markets (Figure 13.3). Sawn products sold locally are mostly used locally too. In addition, the term "Others" in Figure 13.3 includes sales between sawmills, and local demand is more than just "Retail." The Kumano sawmill industry is therefore based on both local resources and local demand. Apart from the wholly retail sawmills, some of the other sawmills also do some retail business.

Figure 13.3

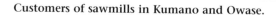

Customers of sawmills in Kumano and Owase.

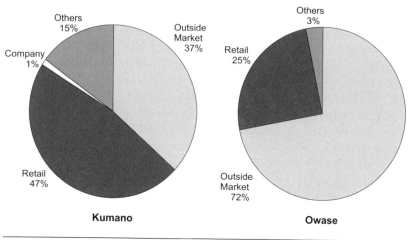

Kumano

Owase

Note: "Outside Market" includes wood product markets and wholesalers. "Others" includes sales between sawmills.
Source: Owase Kumano Ryuiki Ringyo Kasseika Center (1997a, 23).

The second characteristic typical of the Kumano sawmill industry is that market-based sawmills have explored their business options in different ways. As shown in Figure 13.3, timber sold to outside markets represents 37% of the total sales of the Kumano sawmill industry. Each of the market-based sawmills has developed its own line of products and business partners; hence, no particular product is a specialty of the region as a whole. For example, three market-based sawmills interviewed by the author have each developed different businesses. One sawmill specializes in hinoki products and produces first-grade lumber from medium or large logs, mainly first-grade hinoki posts. Another specializes in sugi and produces a variety of products, including both first-grade and second-grade lumber, using medium to large logs. The third one's specialty is again hinoki, but the products differ from those of the first. This sawmill produces first-grade lumber for fixtures, and first-grade veneer for the surface of glue-laminated wood.

In *sanchi,* where many sawmills within an area produce the same specialty, each sawmill can benefit from the brand recognition that their local specialty enjoys, which was originally established by the concentration of sawmills in the region and cooperation among them. They are, however, in competition with each other, since they all sell the same product in the same market. In contrast, the small number of market-based sawmills in

Kumano have individually established both production and distribution of their different products, and compete only peripherally with one another in the same market.

The number of sawmills in the Kumano area has been declining, as has been the case in both Mie Prefecture and the country as a whole. This does not necessarily mean that all of the sawmill enterprises have been declining, but rather reflects the fact that smaller sawmills have gradually closed. According to the survey referred to earlier, which also questioned sawmill owners about the prospects that the future holds for their businesses, some small sawmill owners expect to go out of business in the future, but all of those operating on a larger scale expect to at least maintain the status quo, and hopefully expand their business. Among the owners of the latter type, many were found to have a positive attitude towards their business and were future-oriented. Many of them are second-generation owners in their forties, younger than a typical sawmill owner in other areas, and they have some vitality and new ideas.

The vitality of the sawmill industry in Kumano is reflected in the recent collaborative initiatives of the industry to create new demand for Kumano's timber. Such initiatives include the establishment of a precut timber factory in 1991, which has developed well with increasing demand. Today the factory produces 4,500 m^3 of precut timber per year, about 40% from domestic logs. Another initiative is a factory to produce laminated sugi lumber, founded in 1997. While most other laminated-wood factories use imported wood, this factory is trying to produce laminated lumber mainly from domestic sugi. It will be interesting to see how the business develops in the future.

Factors Supporting the Local Forestry and Sawmill Industries

Although Kumano is an area with a long tradition of forestry, it never developed as a *sanchi* after the war. However, a small-scale system that encompasses the process from log production through to wood consumption has survived in this region. In this section, the factors that have supported the forestry and sawmill industries in this district are examined.

First, we consider the relationship between forestry production and the local sawmill industry. Despite the long-stagnant price of sugi, which is the representative Kumano timber, forestry has remained relatively active in Kumano. One of the factors that has supported forestry production has been the demand from the local sawmill industry. Forestry in this region has also had the advantage of its location on the borders of three prefectures, all of which have active forestry and forest products industries, and the timber from Kumano's forests has also supplied these neighbouring areas. Forestry in Kumano has thus been supported by the demand from both local and neighbouring sawmills.

As for the sawmill industry in Kumano, it has been supported by the abundant forest resources developed in the long history of the area's forestry, and by relatively active forestry production. The small sawmills in Kumano would not have been able to stay in business were it not for these valuable local resources. The major species in Kumano forests is sugi but there are also hinoki trees. Thanks to the traditional forestry that produced large trees before the war, these forests are still more abundant than the ordinary forests in other regions. The timber harvested includes large old sugi and hinoki trees, some of which may be over a hundred years old. Such timber, which is of very good quality and from which first-grade lumber can be produced, is therefore especially important to the market-based sawmills. Needless to say, the rich local forest resources also support the retail-based sawmills, which use a variety of woods to produce the various products necessary for house building. Sawmills in Kumano have the additional advantage of their location, as they can and do procure materials from neighbouring areas.

Second, we consider the sale of sawn products. As noted, retail sales represent a large proportion of the total sales by the domestic sawmills in Kumano. There are some forestry areas like Kumano that have small, retail-based sawmills. Such locally based sawmills were previously found throughout the country, but many have been driven out of business. This was due partly to widespread marketing by the big housing companies, which have continued to expand their business into rural housing markets, since these companies do not procure their materials locally. Urbanization of rural areas has also had an influence. The traditional Japanese post-and-beam construction method generally requires more wood than other methods of construction that are westernized or more industrialized. The more that rural areas become urbanized, the lower the share of traditional wooden housing, and today many small sawmills in rural areas process both domestic and imported wood for economic reasons, although this has not happened in Kumano.

In Kumano retail-based sawmills do business only with domestic wood. Why have these enterprises survived? One contributing factor is the geographically remote location of Kumano. Kumano is a small rural area, situated on the southern end of the Kii Peninsula and distant from major cities, and may therefore not have been an attractive market for the big housing companies. In fact, the area has not been urbanized, and large housing companies have entered the market only very recently. Hence, most houses are still built by local carpenters and small housing enterprises, as was previously the case everywhere, and these carpenters and housing firms use local wood processed by the local retail-based sawmills. Another factor has been the local people's preference for local timber in their housing construction. It is said that most people in the region still believe that local sugi, called

"Kumano sugi," is the best material with which to build their houses, and this is reflected in the share of wooden houses in the total of newly built houses each year: 84% in fiscal year 1996, a much higher figure than the national average of 46% (Ministry of Construction 1999, 56). A similar strong preference by local people for sugi is also found in Kyushu, but it is not very common in other areas of mainland Japan.

As for the market-based sawmills in Kumano, each has made independent efforts in production and marketing. Some enterprises have closed, but those that actively do business today have individually developed different specialties, methods of quality control, and marketing channels. Although they do not enjoy the local brand recognition of a *sanchi*, each enterprise has established its own brand through its own efforts in quality control and marketing. Consequently, both the retail-based and the market-based sawmills have coexisted in the region.

It should be noted, however, that there are negative factors affecting local sawmills of both types. Locally, the population in Kumano has continued to decrease and the retail market has been shrinking, even as big housing companies have gradually expanded their business into this area. Retail-based sawmills now face a shrinking and more competitive local market. Nationally, as the total demand for timber has decreased, so has the demand for first-grade lumber. In addition, laminated lumber is taking market share from first-grade lumber, which is one of the main products of the market-based sawmills. Sawmills of both types in Kumano strongly feel a common and urgent need to find new markets for products that fit the needs of consumers today. This is reflected in the establishment of factories for precut timber and glue-laminated sugi wood, which were financially supported by government subsidies.

In 1991 the government hammered out the political concept of "Forest Management Based on River Basins," to reactivate forestry and to promote sound forest management. Under this concept, the government has promoted each area along the same river basin in order to construct an integrated system, in which forest management, forestry, processing, and distribution are effectively united. Regardless of the new policy, such a system has worked in Kumano on a very small scale, but the forestry and sawmill industries have still declined over a long period. In order to reactivate the Kumano forest sector, the forestry and sawmill industries would need to cooperate more closely, since the local forest resources and the local demand for them are the basis for the forest sector in Kumano.

References

Chiiki Ringyo Kenkyukai, ed. (1978). *Ringyo no Chiikiteki Hatten no Doko to Kadai*. 304 pp. Sozo Shobo, Tokyo (in Japanese).

Forestry Agency (1996). *Ringyo Hakusho,* fiscal 1996 ed. 201 pp. Nihon Ringyo Chosakai, Tokyo (in Japanese).

Handa, Ryoichi (1999). "Tono Hinoki no Shijo Kaihatsu Doryoku." In Chubu Shinrin Kanrikyoku, ed., *Kiso Chiikizai no Hanbai Suishin Hosaku ni tsuite,* 155-68. Nagano (in Japanese).

–, ed. (1986). *Henbousuru Seizaisanchi to Seizaigyo.* 319 pp. Nihon Ringyo Chosakai, Tokyo (in Japanese).

Kumano-shi (1995). *Kumano-shi no Ringyo.* 8 pp. Kumano-shi (in Japanese).

Mie-ken (1988). *Mie-ken Ringyoshi.* 1,001 pp. Ise-shi (in Japanese).

– (1997). *Mie-ken Ringyo Tokeisho,* fiscal 1996 ed., 44-47. Mie-ken (in Japanese).

Ministry of Agriculture, Forestry, and Fisheries (1991a). *Sekai Noringyo Census 1990. Vol. 1. Mie-ken Tokeisho (Ringyohen),* 16-17. Tokyo (in Japanese).

– (1991b). *Sekai Noringyo Census 1990. Vol. 15. Ringyo Sogo Tokei Hokokusho,* 28-29. Tokyo (in Japanese).

– (1996). *Mokuzai Jukyu Hokokusho,* fiscal 1995 ed. 243 pp. Tokyo (in Japanese).

Ministry of Construction (1999). *Kenchiku Tokei Nenpo,* fiscal 1999 ed. 747 pp. Kensetsu Bukka Chosakai, Tokyo (in Japanese).

Morita, M., ed. (1994). *Rinsan Keizaigaku.* 274 pp. Bun-eido Publishing Co., Tokyo (in Japanese).

Owase Kumano Ryuiki Ringyo Kasseika Center (1997a). *Owase Kumano ni okeru Seizaigyo Mokuzaigyo no Doko Chosa Kekka.* 54 pp. (in Japanese).

– (1997b). *Owase Kumano Ryuiki Ringyo Kasseika Jisshi Keikakusho.* 65 pp. (in Japanese).

14
Japan's Wood Trade
Yoshiya Iwai and Kiyoshi Yukutake

Japan's total wood demand was 93,809,000 m³ in 1998. Only 22% of this was produced domestically; the other 78% was imported. By comparison, Japan exported only 200,000 m³ of wood. If Japan imports such a large amount of wood, what is her position in the global trade of forest products?

Table 14.1 shows the five leading wood-importing countries for three different types of wood materials in 1998. Japan ranked first in importation of roundwood and chip and particles, and was second to the United States in importation of sawnwood. When the volumes for the three materials are totalled, Japan ranks first, outstripping the US. In this chapter, we describe how Japan's wood trading developed and how it has changed over the years.

The Liberalization of Wood Imports
Japan's economy was devastated during the Second World War. After the war, government policy emphasized the importation of materials necessary for economic revival in order to speed recovery. With regard to wood, foreign currency was allocated for the importation of lauan roundwood from South-

Table 14.1

Leading wood-importing countries, 1998.

	Roundwood		Sawnwood		Chips and particles	
Rank	Country	Import volume[1]	Country	Import volume[1]	Country	Import volume[1]
1	Japan	15,190	USA	44,940	Japan	26,610
2	Finland	9,328	Japan	7,765	Italy	1,134
3	Sweden	9,300	Italy	7,295	China	1,699
4	China	7,150	UK	7,031	Canada	1,749
5	Canada	6,278	Germany	5,822	South Korea	885
World total		89,329		115,760		37,883

1 In thousands of cubic metres.
Source: Food and Agriculture Organization (FAO) (1998), *FAO Yearbook*, Forest products.

Table 14.2

Changes in wood price index in Japan, 1952-61.

Year	Total average of general prices	Wood price
1952	100.0	100.0
1953	100.4	130.3
1954	99.7	141.8
1955	97.9	127.9
1956	102.2	133.5
1957	105.3	154.2
1958	98.4	149.6
1959	99.4	151.5
1960	101.3	161.1
1961	102.3	194.6

Source: Bank of Japan, *Oroshiuri-bukka-shisuu,* 1952-61.

east Asia, so that plywood processed from the roundwood could be exported, thereby earning more foreign currency. During this time, wood imports were limited to lauan roundwood only.

Japan recovered economically during the Korean War, and experienced a high level of economic growth beginning in the late 1950s. This led to increased wood demand and, as shown in Table 14.2, the price of wood rose rapidly even as the prices of all other products remained generally quite stable. To deal with this situation, trees in the national forest were cut, but the situation did not improve. In 1961 the government was concerned that the price of wood would adversely affect economic growth. It decided to import roundwood and to liberalize trade in sawnwood and plywood by 1964 by reducing import tariffs gradually.

Changes in Japanese Wood Demand and Types of Wood Imported

Table 14.3 shows the changes in the amounts of Japan's wood demand and wood imports, and in the self-sustenance rate. The wood demand was 71,303,000 m³ in 1960, but kept increasing until it exceeded 100 million m³ in the early 1970s. The peak was about 118 million m³ in 1973. Because of economic recession caused by the first oil shock of the 1970s, the demand for wood then decreased until the mid-1980s, before increasing again in the late 1980s.

Imports totalled 7,541,000 m³ in 1960 and grew by eight times in the early 1970s. The self-sustenance rate was 89% in 1960, but decreased sharply so that by 1969 it had dropped below 50%.

As the Japanese economy declined, the amount of wood imports decreased until the mid-1980s. It rose again beginning in the late 1980s; since then the self-sustenance rate has fallen steadily, and by 1996 it had dropped to 21%.

Table 14.3

Changes in wood demand, import volume, and self-sustenance rate of wood in Japan, 1960-98.

Year	Total wood demand (1,000 m³)	Wood import volume (1,000 m³)	Self-sustenance rate (%)
1960	71,303	7,541	89
1963	77,167	16,642	78
1966	82,470	25,041	70
1969	98,385	48,753	50
1972	110,497	63,354	43
1975	99,303	62,190	37
1978	106,344	71,365	33
1981	94,586	60,603	36
1984	93,963	58,772	37
1987	105,382	72,212	31

Note: Volume calculated in terms of roundwood volume.
Source: Forestry Agency (various years), *Ringyo tokei-yoran.*

Table 14.4

Changes in the volume of Japanese wood imports, 1960-99.

Import volume (1,000 m³)

Year	Roundwood	Sawnwood	Chips	Wood-based panels
1960	6,223	156	–	–
1963	13,030	952	9	–
1966	20,735	1,214	503	–
1969	33,741	2,066	4,115	63
1972	42,339	2,422	8,076	170
1975	35,650	2,612	11,340	165
1978	42,653	3,857	13,116	85
1981	29,220	3,898	12,508	191
1984	28,404	4,493	12,156	437
1987	32,292	7,397	14,026	2,444
1990	28,999	9,082	19,043	4,066
1993	23,438	10,622	22,518	5,518
1996	21,336	12,281	26,445	7,463
1998	15,190	7,765	25,721	5,538
1999	16,551	9,740	25,295	6,034

Source: Forestry Agency (various years), *Ringyo tokei-yoran;* Food and Agriculture Organization (FAO), *FAO Yearbook,* Forest products.

Table 14.4 shows the changes in the amounts of different imported wood materials from 1960 to 1999. Roundwood consistently dominated the imports until the early 1990s. Amounts of imported roundwood increased until the late 1970s, then became constant before decreasing in the late 1990s. The amount of imported sawnwood has also decreased in the late 1990s. In contrast, chips and wood-based panels have shown a rising trend. In 1996 the amount of chip imports ranked first, followed by roundwood and sawnwood.

We turn next to the countries of origin of the different types of wood imports.

Sources of Roundwood Imports

Hardwood

Lauan roundwood had been imported from the Philippines even before Japan liberalized wood imports. Wood imports from the Philippines rose after the liberalization of wood imports in 1961. Most of the wood was processed into plywood for use as building materials and for production of furniture, for both domestic consumption and export. As the supply of large-diameter lauan logs suitable for plywood became exhausted in the Philippines, Sabah (Malaysia) and Indonesia took up the slack and became the main suppliers of lauan roundwood to Japan in the 1970s (Table 14.5).

In the early 1980s, Indonesia introduced a policy to develop the domestic plywood industry, resulting in a decline of her roundwood exports in 1982. In 1985 she prohibited roundwood exports, leaving Malaysia as the only source of lauan roundwood imports. Malaysia, too, began promoting her own wood-processing industry and also introduced policies for preserving her natural resources, resulting in reduced logging and prohibition of roundwood exports beginning in 1993. In response, Japan increased its importation of tropical hardwoods from Papua New Guinea.

Softwood

Since the 1960s the amount of wood imported from the US and Russia has been enormous. Hemlock used to be the most imported roundwood from the US, but today that position is occupied by Douglas-fir (*Pseudotsuga* mensiesii). Canada prohibits roundwood exports in principle, and although imports from Canada increased at one time, they have remained relatively low.

Hemlock roundwood was sawn and processed in Japan and used for posts, sills, and interior materials for building houses, competing with Japanese cedar (sugi). Because building materials made of hemlock gradually began to be sawn and processed in the US, the importation of hemlock logs declined and Douglas-fir became the major roundwood import.

Japanese domestic pine was usually used for beams and crossbeams for houses. The amount of pine harvested in Japan decreased, however, because of insect damage; in addition, arch-shaped pine (Japanese red pine, *Pinus densiflora*), which was abundant in Japan and sturdy enough for housing

Table 14.5

Change in volume of roundwood imported by Japan from different countries, 1960-99.

Roundwood import volume (1,000 m^3)

Year	Philippines	Indonesia	Malaysia	Papua New Guinea	USA	Canada	Russia	New Zealand
1960	3,467	12	22	–	378	22	920	147
1963	5,464	22	2,082	–	2,405	257	1,857	230
1965	5,604	79	3,057	–	3,594	642	2,590	401
1969	8,147	2,060	5,279	–	7,955	233	6,039	1,694
1972	5,133	7,010	6,275	–	10,430	252	7,817	1,773
1975	2,998	6,199	6,250	–	9,359	182	7,769	451
1978	1,805	9,216	10,517	–	10,325	311	8,834	812
1981	1,467	4,505	8,370	–	7,401	323	5,647	497
1984	1,011	1,466	9,770	–	7,200	1,193	5,786	306
1987	42	0	12,331	–	9,702	1,898	6,124	389
1990	24	0	10,311	–	10,335	515	4,864	1,343
1993	2	0	5,455	–	7,699	489	5,261	1,721
1996	0	73	3,460	1,741	6,916	119	5,421	2,135
1999	0	61	2,250	943	3,962	837	6,061	1,609

Source: Forestry Agency (various years), *Ringyo tokei-yoran*; Japan Wood-Products Information and Research Center (June 1998, August 2000), *Mokuzai joho.*

beams and crossbeams, was disliked by young carpenters because of the difficulty of the building technique. Douglas-fir from the US became a substitute for Japanese pine. Because beams and crossbeams have various standards for length and width depending on the size and form of the house, they were inappropriate for mass production by American sawmills, so the sawing had to be done by sawmills in Japan according to orders, further increasing Douglas-fir roundwood imports.

The major log species from Russia are spruce (*Picea jezoensis*), Sakhalin fir (*Abies sachalinensis*), and larch (*Larix gemelinii*). Previously they were used as materials for civil engineering and packaging, but today their main uses are for building materials and plywood processing. Because of changes in the former Soviet Union, there was much confusion over harvesting and imports from Russia declined in the early 1990s, although they exceeded imports from the US in 1997 and 1999.

The increasing importation of roundwood from Russia and New Zealand in recent years has been due to the decline of lauan roundwood imports for plywood as the Japanese plywood industry has shifted to producing softwood plywood.

Sources of Sawnwood Imports

Most imported sawnwood is produced from softwood and used for housing materials in Japan. Sawnwood is dominated by imports from Canada and

Table 14.6

Changes in volume of sawnwood imported by Japan from different countries, 1960-99.

| Year | Sawnwood import volume (1,000 m³) | | | | | |
	Canada	USA	Malaysia	Sweden	Finland	Russia
1960	4	149	0	–	–	0
1963	622	277	0	–	–	0
1965	477	269	0	–	–	45
1969	947	646	90	–	–	111
1972	968	873	98	–	–	104
1975	990	1,094	37	–	–	103
1978	1,815	981	161	–	–	127
1981	1,822	1,130	112	–	–	123
1984	1,937	1,328	50	–	–	147
1987	2,787	2,404	338	–	–	181
1990	3,717	2,795	680	–	–	267
1993	5,472	2,301	728	80	130	287
1996	6,064	2,203	507	350	450	406
1999	4,555	822	382	587	673	459

Source: Forestry Agency (various years), *Ringyo tokei-yoran;* Japan Wood-Products Information and Research Center (June 1998, August 2000), *Mokuzai joho.*

the US (Table 14.6), with Canadian imports consistently surpassing those from the US. Sawmills in both countries competed with each other for exports to Japan until the 1990s, when declining old-growth forests on the US west coast and preservation of spotted-owl habitat resulted in a large drop in the American harvest, mainly from national forests. The increase in the price of sawnwood from the US in 1993 (Table 14.6) further reduced American sawnwood exports to Japan. As imports from the US declined, imports from Canada and Scandinavia increased. This situation will be described in the last section, which describes the increase in sawnwood imports from Europe.

Some sawmills were established by mergers of Japanese and Russian enterprises in the late 1980s. As a result, sawnwood exports to Japan from Russia has been gradually increasing, although not in large quantity.

Chips

Table 14.7 summarizes chip imports from various countries. The US is the largest supplier of imported chips, followed by Australia. Australia, however, has rapidly increased its exports and, as of 1998, ranks the same as the US. Chile and South Africa compete for third place.

Most of the chips from the US were softwood until the mid-1980s. Since then, hardwood chips have increased and now comprise two-thirds of Japanese imports. Chips from Australia, Chile, and South Africa are also mostly hardwood. Imports from Canada exceeded 2 million m^3 in 1990, but have decreased since then. Recently, imports of eucalyptus chips harvested from short-rotation forests in Australia, Chile, and South Africa have been increasing.

Table 14.7

Changes in volume of chips imported by Japan from different countries, 1972-98.

Year	Chip import volume (1,000 m^3)				
	USA	Canada	Australia	Chile	South Africa
1972	5,994	199	935	0	0
1975	7,005	0	2,001	0	28
1978	6,945	62	2,963	0	35
1981	5,294	100	3,454	0	53
1984	4,221	1,119	3,975	0	65
1987	5,426	1,185	4,398	122	67
1990	8,154	2,178	4,578	2,432	552
1993	7,999	1,159	5,520	3,211	1,168
1996	9,083	632	6,646	2,847	1,824
1998	7,852	722	7,566	2,441	2,344

Source: Forestry Agency (various years), *Ringyo tokei-yoran.*

Domestic chips account for 30% of pulp materials in Japan, and imported chips for 70%. Most of the domestic chips come from sawmills sawing imported roundwood. In general, domestic chips are from softwood, whereas imported chips are from hardwood.

Factors Promoting Wood Imports

Despite rich forest resources, Japan's domestic wood production has decreased and the country has imported much wood since 1961. There are various reasons for the rise in wood imports. Two important factors were government import policy and the behaviour of large trading companies, called Sogo-Shosha.

The Japanese government equipped the main harbours with the necessary facilities for log importing and processing, such as wharves, sea-stockyards, and factory lots, before large sawmills were built. Harbours for Russian roundwood were built on the Japan Sea shore, and for US and Canadian roundwood on the Pacific Ocean shore. The sawmills were much bigger and had more modern machinery than the old sawmills processing domestic roundwood.

The importation of roundwood was managed mainly by big trading companies, which purchased it in foreign countries, transported it to Japan by private ships, and distributed it to big sawmills in a sustainable way. Compared with domestic wood, imported wood had three characteristics: low price, stable quality, and stable supply system. Let's look at each of these factors in detail.

Table 14.8 shows the changes in the price of posts made from sugi (Japanese cedar) and hemlock used in building houses in Japan. Hemlock is found mostly in forests on the west coast of North America, and is widely used as a substitute for sugi. For 30 years, from the 1960s to the end of the 1980s, the cost of hemlock was about 20% or 30% below that of sugi. This was because sugi is harvested from man-made forest, whereas hemlock is harvested from natural forest; the silviculture cost for hemlock is lower than for sugi. As large machines are used for clear-cutting large areas in North America, the cost of logging could be kept low. Moreover, the price of sawnwood could also be kept low because of mass production by large sawmills, and the cost of transportation to Japan per cubic metre was no different from the cost of transportation from Kyushu (Southern Island) to Tokyo within Japan. Prices began rising in the early 1990s due to restrictions on the amount of wood cut in old-growth forests and pressure from wood demand in the US. This is also the greatest factor in increased sawnwood imports from Europe.

Imported wood was also characterized by stable quality and stable supply system. All the major exporting countries, such as the US, Canada, Southeast Asian countries, and Russia, have extensive forests, and sorting is done according to the diameter and grade of the roundwood cut there. Roundwood

Table 14.8

Comparison of sawnwood prices in Japan, 1963-99.

Price (¥/m³)

Year	Sugi pillars	Hemlock pillars
1963	21,000	–
1966	22,567	18,608
1969	29,583	23,108
1972	39,700	30,600
1975	53,100	38,200
1978	52,100	39,300
1981	52,400	45,200
1984	48,100	42,300
1987	50,400	42,400
1990	51,500	49,300
1993	53,600	54,500
1996	50,600	50,900
1999	44,700	46,500

Source: Forestry Agency (various years), *Ringyo tokei-yoran.*

of the lowest grade is used for pulp materials, while that of the highest grade is exported to Japan. This means that a large quantity of roundwood of similar quality has to be assembled from the extensive forests.

The same is true for sawnwood in North America. Each local sawmill is incredibly large and has a large production capacity, so it is easy to process a large amount of sawnwood of nearly the same quality. The highest grade wood is sometimes selected by local North American manufacturers, but it is generally done either directly or indirectly by the big Japanese trading companies, which ship the wood to Japan in big cargo ships.

The trading companies that import the wood are affiliated with large sawmills or wholesalers in Japan and have expanded the market by selling wood on long-term credit.

Wood Imports in the 1990s

This section discusses three major changes in wood imports in the 1990s.

Increase in Roundwood and Decrease in Processed Wood

The biggest change in this decade was the increase in the share of imported roundwood and decrease in the share of imported processed or manufactured wood. Table 14.9 shows the changes in roundwood and sawnwood imports from 1992 to 1999. During this period, roundwood decreased by 36%, from 25,878,000 m³ to 16,551,000 m³. On the other hand, sawnwood increased by 8%, from 9,047,000 m³ to 9,740,000 m³.

Table 14.9

Changes in volume of roundwood and sawnwood imported by Japan, 1992-99.

| | Import volume (1,000 m³) | |
Year	Roundwood	Sawnwood
1992	25,878	9,047
1993	23,438	10,622
1994	22,386	10,758
1995	21,944	11,807
1996	21,336	12,281
1997	20,407	12,590
1998	15,190	7,765
1999	16,551	9,740

Source: Ministry of Finance, *Trading Statistics, 1992-99.*

Table 14.10 shows changes in processed wood except sawnwood from 1992 to 1999. During this period, plywood increased 1.6 times, MDF (medium-density fibreboard) 3.2 times, particleboard 3.0 times, laminated lumber 12.9 times, and wooden furniture and its parts 2.8 times.

The increase in processed wood imports was generally caused by the strong yen, which made it advantageous to import more value-added wood. Housing companies in Japan increased their importation of processed wood from sawmills or processing mills in foreign countries instead of purchasing it from domestic mills processing mainly imported logs. Some of them established their own sawmills in foreign countries and imported wood materials directly from them.

Table 14.10

Changes in amount of processed wood imported by Japan, 1992-99.

Year	Plywood (1,000 m³)	MDF (tons)	Particleboard (1,000 m³)	Laminated lumber (1,000 m³)	Wooden furniture and parts (tons)
1992	2,985	96,714	125	21	122
1993	4,087	161,314	179	59	145
1994	4,045	250,811	347	89	199
1995	4,399	243,461	361	148	252
1996	5,314	296,019	–	319	–
1997	5,326	339,503	673	289	306
1998	3,872	251,855	411	149	283
1999	4,801	308,907	380	271	347

Source: Japan Wood-Products Information and Research Center (July 1998, August 2000), *Mokuzai joho.*

Diversification of Countries of Origin

The second biggest change was the diversification of countries of origin for wood imports. Hardwood logs imports from Africa have grown since 1990, from almost no imports in the 1980s. Imports from Papua New Guinea and the Solomon Islands replaced Malaysian logs. In general, the diversification was caused by changes on the forest resources of tropical regions.

New countries appeared on the list of import sources in the 1990s: Portugal, South Korea, Italy, and Spain for MDF; Belgium, Germany, South Korea, Portugal, and Thailand for particleboard; and Canada, Sweden, New Zealand, and Finland for laminated lumber.

Increase in Sawnwood Imports from Europe

The third change was an increase in sawnwood imports from Europe, which went up 19 times from 1992 to 1999 (Table 14.11). The major countries of origin were Finland, Sweden, and Austria, which together accounted for 90% of all imports from Europe. Each of these countries has its own active forestry and forest industry. In 1999 imports from Europe reached 1,901,000 m³, putting Europe in second place after Canada in sawnwood imports (Table 14.6).

Sawnwood imports from Europe increased for several reasons. Kiln-dried sawnwood and laminated lumber were needed for housing materials in Japan, and European wood was the most suitable. This new need was created by three factors:

1 The first was a change in the traditional house-building method. In 1999, 1,200,000 units were built, of which 55% were made of wood. Eighty-

Table 14.11

Changes in volume of sawnwood imported by Japan from European countries, 1992-99.

Year	Sawnwood import volume (1,000 m³)				
	Total	Finland	Sweden	Austria	Others
1992	10	2	1	–	7
1993	260	128	82	5	45
1994	591	215	204	78	94
1995	866	366	209	157	134
1996	1,234	448	350	285	151
1997	2,108	588	809	465	246
1998	1,123	421	304	287	111
1999	1,901	673	587	264	177

Source: Japan Wood-Products Information and Research Center (August 2000), *Mokuzai joho.*

one percent of wooden houses were built by the traditional method, in which ground stills are placed horizontally on the foundation and posts set up vertically on the ground stills. These structural materials are connected together by joints or connections made by professional carpenters with chisels and hammers. Because this system required a complex technique and much processing time, costs were higher than with newer housing construction methods using 2 × 4s or prefabricated materials. In the mid-1980s the "precut" system was developed, in which joints and connections were processed automatically by computerized machines in factories. The number of precut factories rose from 150 in 1985 to 900 in 1998, and it is now assumed that the precut system is used in half the houses that would have been built using the traditional method. However, the wood materials processed using the precut system must be kiln-dried to protect them from transformation. Wood with about 20% of water content ratio is needed for building materials.

2   The increased demand for kiln-dried European wood was also one of the consequences of the strong Han-Shin earthquake in 1995. Many wooden houses built after the Second World War were destroyed and the weakness of houses built by the traditional method was apparent. Housing companies building wooden houses needed to make them earthquake-resistant, and increased their use of kiln-dried wooden materials for such houses.

3   The third factor was the quality of houses. In Japan green wood had been used for houses built by traditional method. This had caused many problems, such as doors not closing or floors becoming inclined because of wood warp. In 2000 a law protecting consumers from such problems was passed. Many housing companies began using kiln-dried wood to avoid these problems. Kiln-drying was not widely used in Japan because of its high cost, and the same was true for sawnwood imported from Canada and the US. Kiln-dried sawnwood from North America comprised less than 30% of sawnwood imports in 1998.

In 1993 the cost of sawnwood imported from the US rose due to regulation of cutting in old-growth forests. Japanese trading companies began to import inexpensive, kiln-dried sawnwood from northern Europe. Because Finland and Sweden had been exporting wood materials to Central Europe that were completely kiln-dried, they were able to meet Japan's need in terms of kiln-drying techniques and cost. Imports from Austria have also increased recently (Table 14.11).

The major species imported from Europe is spruce, which is harvested from man-made forests 80-100 years old. Spruce is easy to process and the wood quality is stable because of the old age of the trees, so European spruce is now very popular in Japan.

Since 1995 the demand for laminated lumber as a substitute for kiln-dried sawnwood has increased. Much sawnwood is imported from Europe as materials for making laminated lumber.

To sum up, the changes in Japanese wood imports in the 1990s were due mainly to three factors: (1) the strength of the yen in currency exchange markets, which created an incentive to import value-added wood; (2) changes in forest resources in the US and Southeast Asia, which led to increased imports from other countries; and (3) changes in the demand for products such as kiln-dried wood.

References

Araya, Akihiko (1998). *Indonesia goban sangyo*, 55-152. Nippon Ringyo Chosakai, Tokyo (in Japanese).

Iwai, Yoshiya (1990). *Nihon no jyutaku-kenchiku to kita-America no rinsangyo*, 1-129. Nippon Ringyo Chosakai, Tokyo (in Japanese).

– (1992). *Youroppa no shinrin to rinsangyo*, 1-138. Nippon Ringyo Chosakai. Tokyo (in Japanese).

– (1994). "Gaizai." In Morita M., ed., *Rinsan Keizaigaku*, 159-80. Bun-eido Publishing Co., Tokyo (in Japanese).

Part 3
New Trends for Forestry in Japan

15
Depopulation and *Mura-Okoshi* (Village Revival)

Takashi Iguchi

Depopulation of rural areas is an old problem, with origins in the 1960s and the recovery of the postwar Japanese economy. After more than 30 years, however, it is also a contemporary problem. The outlook in depopulated areas has changed greatly compared with the situation at the time the phenomenon began. Whichever depopulated area is visited, the gloomy image of the old days is no longer seen. Centrally located villages with splendid new town halls and a whole range of facilities radiate an atmosphere of rosy optimism. In spite of all this, however, the depopulation problem is steadily worsening, while an already aged population continues to grow older. As integrated elementary and junior high schools are concentrated in the vicinity of town halls, it is still possible to hear the voices of children there; but move away from the town halls into more remote villages and it becomes difficult to find any sign of human activity, even in the daytime. When someone is eventually chanced upon, it usually turns out to be an elderly person engaged in his or her solitary chore.

The so-called *Mura-Okoshi,* so popular in recent years, is an expression of the strenuous efforts made in these areas to somehow turn the tide. In contrast to the depopulation measures handed down, as it were, from above, there are projects that have recently become popular throughout Japan by trying to revitalize areas and shaping them into suitable living environments by relying on the resourcefulness and ingenuity of local residents. These projects are referred to as *Mura-Okoshi* ("Village Revival") or "*Mura-Okoshi* Movements." But will these efforts be successful? And will they be able to create a stable livelihood for local inhabitants?

In this chapter, I have tried to present a comprehensive account of the depopulation problem and of *Mura-Okoshi,* which is drawing so much attention as a potential countermeasure. The depopulation problem and *Mura-Okoshi* are a phenomenon and a movement common to agricultural and fishing villages in all regions of Japan, but in this discussion the focus is on mountain villages regarded as typical of the depopulation problem. Actual examples mainly relate to mountain villages in the Chugoku Mountains district in western Japan.

The State of Depopulation

Origin and Mechanism of Depopulation

Before 1960

From 1945 to the late 1950s, the economy of mountain villages was relatively stable and in the process of achieving full potential. This vitality was due to a variety of factors, notably an increase in agricultural productivity through independent farming made possible by the land reforms of 1946, relatively profitable farm produce prices, expanding independent charcoal production, and new income due to increased demand for a wide range of forest products. To a large extent the economy was self-sustaining, making use of the limited local labour force. Local resources of the mountain villages were managed efficiently, especially with regard to a diverse and complex

Figure 15.1

The mountain village economy in Japan before 1960.

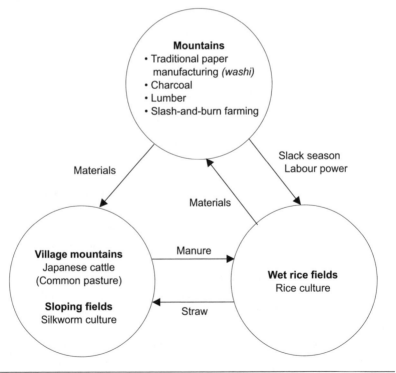

Source: Nagata (1988, 237).

utilization of the farm and forest land. The three products that formed the pillars of the mountain village economy and provided the basic source of income for farmers at this time were rice, Japanese cattle, and charcoal. The three were interrelated and formed a complex economic structure. Nagata (1988) presented this structure as shown in Figure 15.1.

The Causes and Mechanism of Depopulation
As we have seen, the mountain village economy enjoyed a short period of relative stability after the Second World War, and even showed some signs of growth at that time. This situation did not last long, however, as the Japanese national economy entered a period of high growth in the late 1950s. The rapid expansion of urban labour markets in the wake of urban industrialization brought about a rapid decline in mountain village economies.

To begin with, the production of firewood and charcoal, which had always been an important source of income and employment for farmers during the slack winter season, began to decline after peaking in 1957. This decline was due to the drop in demand for charcoal and firewood for domestic use as they were replaced by oil, electricity, and gas. It also deprived the surplus labour force of job opportunities and destabilized the source of income essential to farming household economies. On the one hand, the decline seen in the breeding of Japanese cattle, induced by the simultaneous introduction of farming machinery and the spread of chemical fertilizers, brought about a switch in farming policy from one aimed at fulfilling requirements for diverse and complex management systems to one attempting to raise farming income through a more efficient labour force relying on large-scale management and single-crop farming. These changes caused an outflow of the surplus labour force from mountain villages, a tendency that was advantageous to the labour-hungry urban industries. On the other hand, the diversified forest products industry, which had always been so helpful to mountain village economies, also started to decline, hampered by cheap oil products and imported substitutes. In addition, the "Rice-Field Reduction Policy" was initiated by the government in 1969 to remove the last of the three pillars supporting the mountain village economy, dealing, both materially and morally, a devastating blow to farmers in those villages.

The outflow of population from mountain villages was both massive and constant owing to the effective interplay of the "push-factor" within villages that found it more and more difficult to support their population because of the steady decline of job opportunities, and the "pull-factor" from urban centres. People who left their villages because their labour was not essential in the slack winter season included even the heads of families – the central support of both agricultural and forest industries. This tendency proved to be lasting, as whole families began to leave villages en masse. Thus was the advent of the depopulation phenomenon.

The rate at which population declined between 1960 and 1965 reached an all-time high in Yasaka village in Shimane Prefecture, where a decline in population from 5,288 to 3,446 occurred, representing a 33% loss over a span of five years. Initially, the flow of labour from mountain villages to urban centres proved beneficial not only to urban industry but also to villagers who had always been troubled by lack of job opportunities. The incessant drain in labour seemed to be limitless, however, even as it affected population levels relying on age structures necessary to maintain the living standard and productivity dependent on local resources.

Finally this drain came to threaten the very existence of mountain village society. Large-scale and abrupt population migration caused a deterioration in all local industries and changed the mountain village economic structure. The stagnation of local industries and the decrease in population triggered a financial shortage in local government bodies. When inadequate countermeasures failed to ease these problems, both population and household numbers dropped further. The gravest consequences, however, were the loss of community spirit affecting whole regions and the total collapse of some local communities. Communal work in the village, always regarded as a standard unit in daily life, disappeared, making it all the more difficult for the remaining villagers to maintain a normal livelihood and production level. In some cases, whole villages were forced to move to other areas. Moreover, elementary and junior high schools were integrated or closed, as a smaller population also meant fewer children. The loss of local schools meant the loss of the spiritual foundation necessary for permanent settlement of community members, and it initiated a vicious circle that, again, led to a decline in population and families. Finally, the life force of whole regions started ebbing away from outlying villages until they totally disintegrated.

The foremost authority in the field of depopulation studies, Adachi (1968), summarized the process and mechanism of depopulation as shown in Figure 15.2.

Definition of Depopulation

What is depopulation? To what conditions does the term refer? Its definition when used in a general sense is found in the *Report of the Local Sectional Meeting of the Economic Council* published in 1967: "A situation where, due to a decrease in population, it has become difficult to maintain a normal level of living, to maintain basic public services for the local community (e.g., disaster prevention, education and health care), and where, owing to the inability to rationally exploit resources, local productive functions suffer a remarkable setback." It is obvious that this definition is not satisfactory given the depopulation mechanism expounded above. Adachi (1968) adds to this definition the reciprocal action of the adverse effect upon village maintenance and community spirit caused by the decrease in households.

Figure 15.2

Process and mechanism of depopulation in mountain villages in Japan.

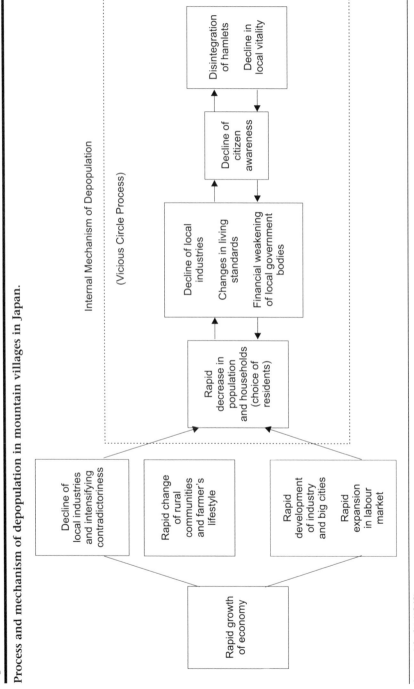

Source: Adachi (1968).

Naito (1991) similarly attaches great importance to community spirit and the village community as the fundamental unit of a region, supporting the amplified definition that includes communal functions and community spirit.

Current Status of Depopulation

Continuing Population Decline
When we look at population movement in depopulated areas (Figure 15.3; National Land Agency [1996]), a high rate of migration can be observed (over 12%) during the 1960s; the situation improved drastically in the 1970s, however, and the improvement continued into the late 1980s. This implies

Figure 15.3

Fluctuation in rates of population increase/decrease in depopulated areas, the three largest city spheres, and regional spheres in Japan over the period 1960-95.

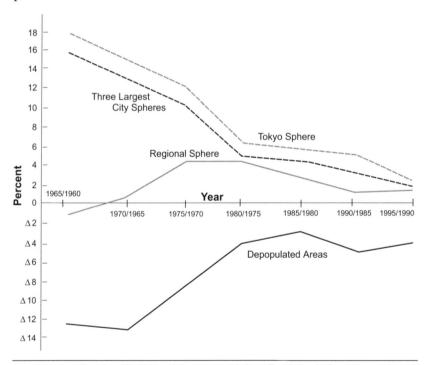

Note: Figures are based on the National Census, except those for 1995, which are the provisional data released for the census that year. The three largest city spheres are: (1) the Tokyo Sphere (including Saitama, Chiba, and Kanagawa prefectures in addition to the Tokyo Metropolis), (2) the Osaka Sphere (including Kyoto, Osaka, and Hyogo Prefecture), and (3) the Nagoya Sphere (including Aichi and Mie Prefecture). The regional spheres include all regions outside these three city spheres.

that the various measures taken to counter depopulation and promote the development of mountain villages effectively reduced the rate of depopulation. It is true that the so-called U-turn and J-turn phenomena (the term "U-turn" denotes the return of young people who had left for the city to their native village; "J-turn" denotes their return to a local city close to their hometown or village) of the late 1980s were marked, and that in some areas, some villages actually saw their populations rise once again. Subsequent developments cruelly betrayed these expectations, however. When the results of the National Census of 1990 were released, it became clear that the population decrease had once more taken a turn for the worse in the latter half of the 1980s. In addition, figures from the 1995 National Census indicate that although the rate had dropped a little, it still remained as high as 4.7%.

Simultaneous Advance of "Social" and "Natural" Population Decline
Again, when looking at population shifts over an entire area, it is evident that although the problem of social decline continues to exist, the situation has shown some improvement lately. On the other hand, in terms of natural transition, annual death rates began overtaking birth rates in depopulated areas in 1987, and as a result the population fell into a natural decline pattern (Figure 15.4; National Land Agency [1996]). Depopulated areas must now reckon not only with social decline resulting from excessive population outflow but also with concurring natural decline. Many communities, particularly remote villages, face the threat of extinction. Needless to say these areas will find it very difficult to maintain, let alone increase, population.

Progressively Aging Population and Increase in Elderly Households
The population decline in depopulated areas is not simply a quantitative decrease across the whole population but implies a clear decline in the younger stratum of the community. The aging of the local population is rapidly increasing, against the background of a steady outflow of people and with almost no hope of successors or heirs to make the U-turn or J-turn. The result is an increase in elderly-couple or elderly single-person households. In 1995 the proportion of households consisting of a single elderly person in depopulated areas was 9.6%, almost twice the national 5.0%. The nearly equivalent percentages for elderly-couple households were 9.3% and 4.6%, respectively. How much longer will these outlying hamlets, with such a high proportion of elderly households, be able to survive? They are obviously in an extremely precarious situation.

Inadequate Management of Farmland and Forests
The diverse potential of farmland and forests can be realized only through proper management. To maintain management, however, it is imperative that mountain village society, especially villages and families occupied in

Figure 15.4

Natural population fluctuations in depopulated areas and nationwide in Japan, 1970-94.

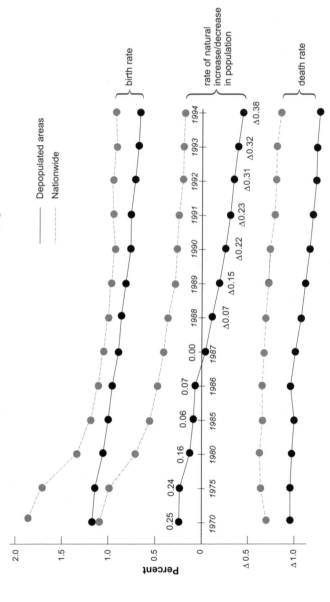

Notes: (1) Figures for 1970 and 1975 are based on vital statistics for those years; other figures are based on population movement tables and national population/household tables derived from basic residential registers. (2) Figures for 1970 do not include Okinawa. (3) Figures for depopulated areas for the years 1970-84 do not include the seven groups mentioned in the additional public notice of 1 April 1986. (4) The natural fluctuation rate for 1987 was 0.04.

agriculture and forestry, be supported appropriately and be allowed to enjoy economic stability. In relation to forestry, woodsmen, who have traditionally been mountain village dwellers, are now themselves steadily aging and on the decline, with the advance of depopulation and the aging of their communities.

Recently, some attempts have been made to recruit a younger workforce. For instance, a "Third Sector" (organizations operating with capital jointly provided by local government bodies and private enterprise) has been established. These efforts have been only partially successful, however, and the industry is still confronted by an extremely severe labour shortage. As a result, essential forest management has been lacking and abandoned woodlands remain in a state of utter neglect. The number of forests no longer capable of fulfilling their expected environmental functions is increasing. The same can be said of farmland. Terrace fields in mountain areas are increasingly being abandoned, and the environmental functions of cultivated farmland are deteriorating due to poor management and work practices.

Depopulation Countermeasures and Their Effects

Legislative Countermeasures

The Depopulation Acts
In order to cope with this whole range of worsening problems, some independent measures were taken by individual prefectures with depopulated areas. Shimane Prefecture, for instance, started the "Rapid Population Decrease Areas Comprehensive Countermeasure Projects." In addition, a movement was started, mainly by the Shimane, Kagoshima, and Kochi prefectures, to put pressure on the national government to introduce special legislation. This movement stirred the government, which had failed to formulate any concrete depopulation countermeasures, into action. The Act on Emergency Countermeasures for Depopulated Areas was established in 1970. The Act on Special Measures for Development of Depopulated Areas was introduced in 1980 and was amended in 1990 to become the Act on Special Measures for Depopulated Areas; it is still in force today. These acts are generally known as the Depopulation Acts. Areas are designated as "depopulated" under the Depopulation Acts using the current city – town – village units on the basis of a population factor (rate of population decrease) and a financial factor (financial power index).

Depopulated areas represent 1,208 cities, towns, and villages nationwide – about one-third of all municipalities (as of 1 April 1996). In Shimane, roughly two-thirds (64.4%) of all municipalities were designated as depopulated areas (Figure 15.5). The only exceptions were 8 cities along the Japan Sea coast and 12 towns and villages functioning as suburbs for these cities.

Figure 15.5

Distribution of depopulated municipalities in Shimane Prefecture as of 1 April 1995.

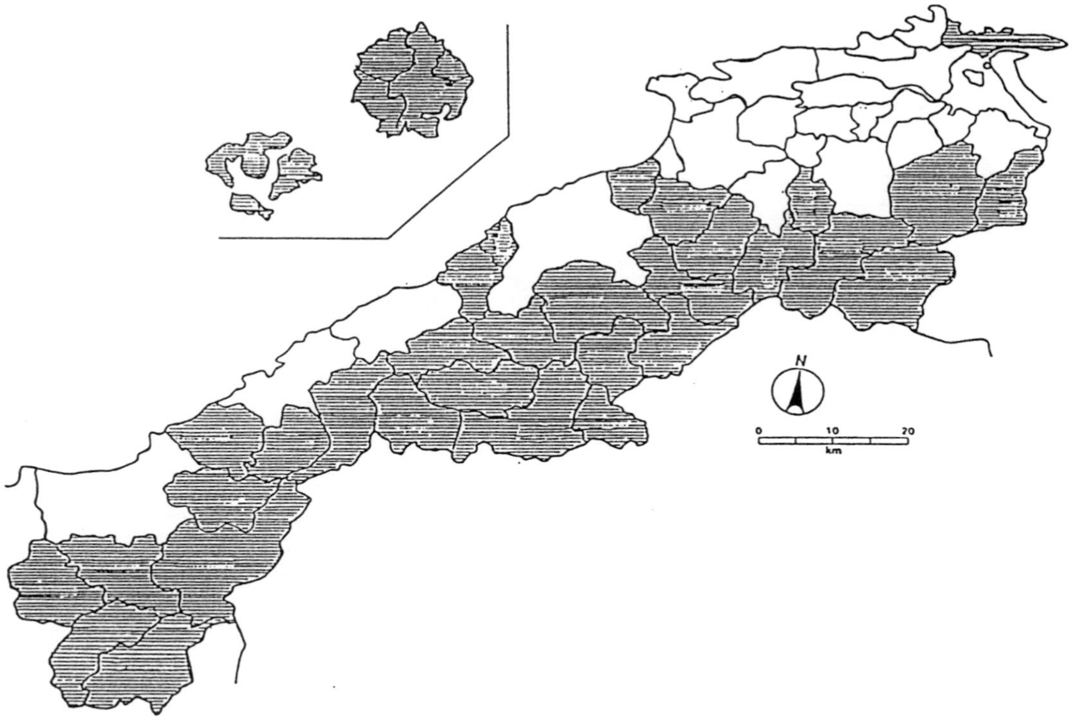

Countermeasures Based on the Depopulation Acts
The wide variety of countermeasures relating to depopulated areas range from streamlining of traffic and communication networks, and of educational, cultural, and living environment facilities, to the promotion of industrial activities. Methods used by the government to promote these measures consist mainly of raising subsidy levels, favourable treatment with respect to taxation and financing, and the floating of so-called Depopulation-Bonds (short for "Bonds for Depopulation Countermeasure-Projects" – special local bonds covering the cost of projects planned and carried out by depopulated municipalities, where 70% of the principal and interest are subsidized by government funds).

Results Achieved by Countermeasures Based on the Depopulation Acts
Over the years, various countermeasures emphasizing modernization and upkeep of roads and facilities have been implemented in mountain villages experiencing ever-worsening depopulation problems. Total investment during the 20 years up to 1989 amounted to about ¥25 trillion, with half spent on the upkeep of traffic and communication networks, with emphasis on roads. One result of this policy has been that, although they may still be compared unfavourably with those in urban areas, living conditions in mountain villages have improved markedly. In some cases outlying hamlets have not been affected by these changes, but in centrally located villages the modernization of roads and various public facilities has in large measure been accomplished. As far as so-called "hard" infrastructure is concerned, the reorganization has attained a relatively high level, clearly a result of the countermeasures.

In reality however, as has been illustrated, the depopulation problem appears to be beyond solution, and is even worsening in certain areas. This suggests that depopulation cannot be solved only by reorganizing "hardware." It is from this perspective that we need to consider the appearance in recent Depopulation Acts of terminology such as *Mura-Okoshi* and other terms indicating promotion of joint management ventures.

This discussion has been limited to the Depopulation Acts, although many municipalities in depopulated areas have received a special designation within the framework of other related acts on local development. A great variety of promotional development projects are also being carried out under these related acts.

Depopulation Countermeasures to Attract Manufacturing Industry
Earlier measures to develop depopulated areas were aimed at encouraging manufacturing industry in an attempt to halt the outflow of labour. Indeed, there was a time when it seemed that this policy would be effective, with newly located factories beginning to absorb part-time labourers and even

succeeding in inducing young people to make the U-turn back to their home-town (village). As has been variously pointed out, however, the main fac-tors that led to businesses moving to rural areas were the availability of cheap labour, good-quality resources, and/or a suitable environment. Con-sequently, this policy of attracting industry, even though it meant creating job opportunities locally so that labourers no longer had to move to other parts of the country, was unsuccessful in providing the type of work desired by young professionals with a high level of education.

Today the idea of creating employment by encouraging manufacturing factories to move to rural areas is regarded as a failure. The aging of the population has made it difficult to procure an adequate labour force, thereby endangering the continuing operation of some factories. On the other hand, rationalization and restructuring have brought about a more integrated working environment in which conditions are now different from the origi-nal ones. These changes sometimes cause workers to develop an aversion to their work. In any case, the policy has lost its relevance as a means of creat-ing new employment.

Mura-Okoshi Projects

Mura-Okoshi projects are so varied that it is difficult to categorize them with a single term. A common feature of successful attempts is that they have been set up and are being carried out voluntarily and spontaneously, rely-ing on the intrinsic, latent energy existing in an area. The ultimate goal is regional development and stimulation of regional activities. As far as the actual content of these projects is concerned, they may consist merely of holding events, establishing exchange programs, or creative manufacturing or processing of local specialties, as in the "One Village – One Product" movement. The underlying concept is for the people of a depopulated area to work together towards a common goal – something that might be called the "Dare-to-Do Revolution" – thereby returning pride to the local village and self-confidence to its people, while at the same time raising a spirit of solidarity. Since the first half of the 1980s, *Mura-Okoshi* movements have started up spontaneously all over the country. Some have ended in failure but many more have continued and been successful. This is why the impor-tance of *Mura-Okoshi* movements has gradually become recognized, in that they lead to an uplifting of citizen awareness and offer an opportunity to reverse the depopulation process. These movements cannot be fully effec-tive, however, by relying only on the efforts of mountain village communi-ties. As will be seen presently, continuous social exchange and conducting of various activities in concert with citizens of urban areas are equally im-portant. It is worth noting that the most successful cases have also been successful in this last respect.

Figure 15.6

Numbers of interchange projects with urban centres initiated in depopulated areas of Japan, 1993 and 1995.

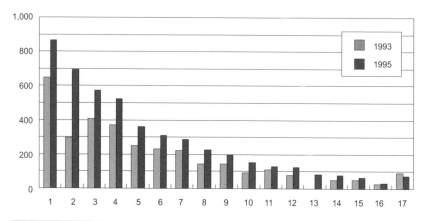

The data are based on investigations by the National Land Agency in October 1993 and October 1995, and have been condensed on the basis of multiple answers (no 1993 data for number 13). The projects numbered 1 to 17 were as follows:

1 *Holding of events.* Attempts to encourage local revitalization through public relations and other image-improving activities such as village and other festivals that rely on the participation of city residents.

2 *Furusato (native village) membership societies.* These societies offer membership to city residents who have no rural roots and others who have left their villages for the city. They attempt to link city and village and promote a return to the countryside through participation in village festivals and by sending seasonal local delicacies and information to members.

3 *Morning fairs, open-air markets, local specialties exhibitions.* Projects that try to appeal to city residents by selling farm produce and other products at morning fairs and open-air markets. Occasionally, exhibitions of products typical of the area are held in the city.

4 *International exchange.* Exchange projects that satisfy mutual needs while fostering new understanding of the local culture and history of a specific locality and their residents abroad.

5 *Forming sister city affiliations.* Various exchange programs that attempt to satisfy mutual needs by linking up with a specific city and its residents.

6 *Sports events exchanges.* Attempts to establish links with city residents through sports, leading to mutual understanding and goodwill. Full use is made of the sports facilities and exchange programs providing local accommodation.

7 *Personal-experience farming, tourist farming.* Allowing city residents, especially schoolchildren, to experience village life and farm work or enjoy the pleasures of farming in a tourist program on local farms and farmland.

8 *Try-the-country-life tours.* Tours aimed at offering city residents a chance to experience simple life in a rural village or be actually productive in agriculture or forestry (snow-shovelling tours, rice-planting and harvesting tours, etc.).

9 *Hosting of nature classes.* Organization of various classes specializing in outdoor life activities and outdoor schools that make optimal use of the rich natural surroundings of rural villages. These projects also include short-term student exchange programs in mountain villages and experience-based training for families with children.

10 *Providing information regarding vacant houses.* Offering information to city residents concerning houses left vacant due to depopulation, with the aim of promoting permanent settlement.

▶

◄ *Figure 15.6*

11 *The Ownership System.* Program whereby city residents provide the necessary capital to own fruit trees or livestock. Management is taken care of locally but the fruit may be claimed by the owner at harvest time.

12 *Hosting of School Excursion Parties.* Programs that aim to establish close links with a specific school by organizing school excursions in conjunction with farmhouse-inns and providing firsthand experience of rural life and agricultural or forest work.

13 *Setting up of satellite-shops.* Establishing facilities in urban centres that gauge the need for rural specialties and function as public relations centres for rural communities.

14 Hosting of *Exchange Students in Mountain Villages.* Programs that allow elementary school children or junior high school students to spend a leisurely one- or two-year period away from home, with farmer foster parents or at an exchange student centre in a mountain village.

15 *Establishing a mini-state or artists' village.* Attempts to strengthen a community in an atmosphere of parody and humour by creating an independent mini-state in Japan.

16 *Promotion of a second house.* Attempts to increase permanent settlement by encouraging the building of second dwellings on abandoned land or by utilizing vacant houses.

17 *Others.* Projects not mentioned above.

As can be seen in Figure 15.6, a great number of municipalities in depopulated areas maintain an exchange program within the framework of a *Mura-Okoshi* movement. The actual number of these programs is increasing, with 70% of these municipalities being engaged in exchange programs with an urban centre.

Potential and Tasks of *Mura-Okoshi* in Solving the Depopulation Problem

Creating a Mechanism to Revitalize Depopulated Mountain Villages

In order to infuse new life into depopulated mountain villages, it is necessary not only to stop the depopulation process but also to reverse it. To bring about this reversal may be considered the most important task of depopulation measures. So far, however, the fact that depopulation is still advancing, and in some areas has reached very serious proportions, shows that such measures have not succeeded in this task.

On the other hand, *Mura-Okoshi* movements that have been established in such great numbers recently appear to be functioning adequately and to have reversed the depopulation process, triggering a positive cycle in some areas. It is here that the possibility of overcoming depopulation is to be found. It is difficult, of course, to argue that the current *Mura-Okoshi* movements are at a level where they can offer the prospect of successfully revitalizing depopulated areas nationwide. A great variety of case studies have been presented and numerous models analyzed, yet even so-called high-potential cases do not necessarily show bright prospects in all situations (Iguchi 1995). The *Mura-Okoshi* of recent years, however, give the impression of being in the process of aligning internal and external factors that may lead to an improvement of those prospects. One internal factor is the changing awareness of rural residents. Although it is possible that depopulation has reached an all-time low, thus lowering the level of resistance to the inflow of outsiders, now, in fact, newcomers are actually being welcomed in badly depopulated mountain villages. External factors include, first, a growing interest in nature in general, and, more specifically among people living in urban areas, towards forestry and agriculture; and second, a sudden rise in the numbers of people wanting to move to mountain villages or aspiring to careers in agriculture (see, for example, Iguchi et al. [1995]).

The *Present Condition of Depopulation Countermeasures, 1995 Edition* (National Land Agency 1996), the so-called white paper on depopulation, refers specifically to this phenomenon, stating on page 19:

> It is also necessary to pay attention to some hopeful signs, among others, permanent residents generated by "U, J or I-turns" and the increase in people who are linked to the area by an exchange program, both of which have become possible thanks to the way the Japanese have changed their evaluation of nature and the environment in recent years, a change which has led to spontaneous and autonomous local development projects.

As a change independent of local internal or external factors, it shows that a certain percentage of the Japanese population, having experienced the "bubble" after a period of high economic growth, are re-evaluating their lifestyle and trying a new direction. These internal and external changes run parallel to efforts that stimulate Village Revival within the context of an exchange between urban and rural communities. These changes are expected to improve opportunities for revitalizing depopulated areas based upon the Village Revival movement.

Changes in Values

While high economic growth continued, villagers and city dwellers had the same values. Securing higher income was considered of utmost importance. This was an idea shared by the whole nation at that time, and people were ready to make any sacrifice necessary to achieve it. Besides changes in the natural environment, tradition, and landscape of mountain villages, the result of this attitude was depopulation – a shift of people towards the cities, where higher income was available.

Recently, however, it has been recognized that the apparent size of one's income is unrelated to the quality of life. Concerns about environmental problems are ever-increasing. In this context, therefore, people with a non-traditional viewpoint and value system have also come to the foreground in mountain villages. The number of people interested in organic farming and willing to actually engage in it is increasing. This is a period in both the city and mountain villages when people with new values coexist with those who maintain traditional views. It can be said that both homogeneous and heterogeneous values are developing independently of each other. If the *Mura-Okoshi* movement is to survive and be successful, it must include exchange between rural and urban residents with a common set of values. Moreover, these values must differ from traditional values, and emphasize nature and the joys of life more than the rewards of high income. *Mura-Okoshi* will have to be built on exchange, cooperation, and solidarity with people from urban centres who have acquired a new outlook on life.

Tasks of the *Mura-Okoshi* Movement

Investigating New Local Resources and Their Integrated Development
Local resources are usually considered to consist of minerals, forests, farmland, marine products, and other natural resources that can be used as a basis for production and enjoyment. The local resources under consideration here, however, also include the climate; natural features such as the land, sea, mountains, rivers, and lakes; and also the history, historical and scenic sights, culture, technology, and human resources existing in the region. In this sense, they represent all actual and potential characteristics that give a locality a special flavour. It is necessary to identify these resources and develop methods to promote them in a variety of ways. It is important to utilize them, not in a random, unrelated fashion but, as far as possible, integrated into a unified strategy that can mutually benefit primary, secondary, and tertiary industries in the region.

Organizing of Combined Management
The industries most suitable to depopulated rural villages are agriculture and forestry. Further, it is difficult to increase the scale of management under common conditions existing in rural villages. Combined management is one possibility for enlarging operations that initially have to take into account limited farm and forest land. Under combined management, a comprehensive and efficient plan can be implemented to apportion family labour and to utilize farm and forest land. Even if prices of one crop should fall, these losses could be recouped through the sale of another crop, thus ensuring the stability of the farming household economy. Combined management can establish a mutually secure relationship between the livestock and cultivation sections and develop methods of avoiding pollution and soil depletion. As this type of management creates job openings for the aged, it also gives them a purpose in life. In this way, long-term stabilization of agriculture and farm villages becomes a reality.

As a matter of course, such combined management does not lead in the same direction as traditional large-scale policy, which attempts to obtain income equivalent to that of urban areas through mass production and marketing of specified farm and forest products. It is a form of management that encourages settlement and provides a secure livelihood. It involves a rethinking of life and productive work in rural villages based on changes in lifestyle and values. The form of agriculture selected here would probably be along the lines of organic farming, which exploits the full potential of mountain village agriculture.

Interacting with the City as a Support for Mura-Okoshi
A very close correlation exists between successful *Mura-Okoshi* and exchanges between mountain villages and urban areas. Ideally, this exchange should have a reciprocal and continuous character, and be conducted on an equal footing. Only on this basis will a cooperative and mutually supportive re-lationship between residents from both areas be created. A form of exchange with urban centres based on unilateral sacrifice on the part of mountain vil-lages is destined to be short-lived and would lead to the failure of *Mura-Okoshi*.

Whether the production and processing of agricultural and forest prod-ucts relying completely on the use of the local resource potential in the context of a *Mura-Okoshi* project are successful or not depends, for example, on how smoothly marketing (distribution) mechanisms function. For prod-ucts that are not easily marketed through normal channels, new channels must be developed. To achieve this, an exchange between city and village, maintained in one form or other, would be helpful, and by giving substance to this exchange, the success of *Mura-Okoshi* would be assured. This is true not only of *Mura-Okoshi* centred on production, processing, and marketing of products or produce but also of *Mura-Okoshi* relying on the promotion of events.

Training Capable Personnel and Local Leadership
Wherever a *Mura-Okoshi* movement is successful and the region is full of animation and activity, a cadre of able personnel and a local leader invari-ably exist. A local leader must be capable of drawing out the constructive potential of the community and making optimal use of it by providing a direction and integrating its potential. It is therefore necessary to consciously develop human resources that can fulfill this task in cases where there are no (or too few) local leaders.

Discovering the Strength of the Rural Community: The Meaning of Self-Support
Aiming at self-support is a matter of course not only in the case of indi-vidual farms but also at the total local community level. As mentioned, a policy leading towards combined management should be the objective in mountain villages because it would foster self-supporting production and the broadening of the production range. Although the word *self-support* has

various meanings, here it refers to an alternative income and relates to a reassessment of the rich local food culture and the creation of a healthy local environment.

Local self-support can be first considered as an alternative way to produce income. Methods of increasing income are usually thought to involve, on the one hand, a maximum expansion of product sales and, on the other, a maximum reduction in necessary production costs. This is no doubt true when considering only production, but when taking a broader view of the farming household economy or the regional economy overall, consumption has to be taken into account. In other words, self-support should be considered an attempt to maintain the same balance by reducing not only the amount of cash income but also of expenditure. This method of raising the level of self-support is the easier approach for rural communities.

The key to stability and wealth in a farming economy lies in the ratio of its self-support rate. This economy, therefore, differs in character from an industrial economy, where 100% of production is considered merchandise. As a consequence, it is necessary to regard self-sustenance, together with quality and quantity, as extremely important – as being a strength of the agriculture, forestry, and other primary industries.

Showing Pride in, and Attachment to, the Locality
In order to promote settlement in an area, it is imperative that at least the people living in the region already show pride in, and attachment to, their village. This is the precise aim of the *Mura-Okoshi,* and determines success or failure. People from the city will feel attracted only to areas where the local residents themselves show pride and attachment to the village or area.

Conclusion
It is difficult to calculate the value and appeal of rural areas in monetary terms. Thus, ultimately low-income potential is perceived from the outside as the poverty of the countryside. Depopulation countermeasures with origins in this perception attempt to stimulate production in order to raise rural income levels to those of urban areas. When expansion of the production scale proves difficult, an alternative is found in the promotion of a high value-added type of agriculture. In any case, although the latter may offer some possibilities, it is impossible to help rural areas out of their difficulties through this policy. It is utterly impossible for people who make their living in a rural area to achieve the same income as urban residents. They do, however, have the potential to achieve a quality of living different from the city. (Comparing the two is difficult, as quality of living is not quantifiable.)

The attraction of mountain villages cannot be measured financially. To understand this, value systems have to change. Recently some people both in and outside villages have become aware of this. It was Adachi (1979) who once expressed the difference in quality of life as follows: "A million yen in a village is two million in a city." In a village it is possible, with an income half that of a city income, to lead a life far more abundant than is possible in an urban area.

A rich imagination and firm resolve are indispensable for *Mura-Okoshi* to reverse depopulation. Moritomo (1991), however, expresses some reservations when he states: "If we do not pay serious attention to the philosophy of the *Mura-Okoshi* movement, if we do not consider local ways of life and culture and proceed to create a movement that fundamentally affects the life and culture of all Japan by involving the urban sector also, we may say that the efforts made today will prove fruitless." This author agrees. It is necessary to develop a powerful movement so that *Mura-Okoshi,* after all the effort already invested, does not die a slow death.

References

Adachi, Ikutsune (1968). "Kaso towa nanika," *Noosonkaihatsu* (1): 81. Simane Daigaku Noogakubu Noosanson Chiikikaihatsu Kenkyu-Choosasitsu (in Japanese).

– (1979). *Mura no saisei,* 153-79. Nihon Keizai Hyooronsha, Tokyo (in Japanese).

Iguchi, Takashi (1995). "Toshi to noosanson no kooryu." In Kitagawa, Izumi, ed., *Cyuusankanchiiki keiei-ron,* 315-45. Ochanomizu Shobo, Tokyo (in Japanese).

Iguchi, Takashi, Ito, Katsuhisa, and Kitagawa, Izumi (1995). "Cyuusankan chiiki ni okeru nooringyoo-seisan to teizyusokushin-seisaku ni kansuru ikoochoosa no bunseki (I) Cyuusankanchiiki eno izyuu no kanoosei ni kanshite" ("Analysis of agriculture – forestry production and settlement promotion policies in rural areas (I) On the possibility of transfer to rural areas"), *Nihon ringakkai-shi (Journal of the Japanese Forestry Society)* 77(5): 421-28 (in Japanese with English summary).

Moritomo, Yuuichi (1991). *Naihatsuteki hatten no michi,* 51. Noobunkyoo, Tokyo (in Japanese).

Nagata, Keijuro (1988). *Chiikishigen no kokuminteki riyo,* 237. Nosan Gyoson Bunka Kyokai, Tokyo (in Japanese).

Naito, Seichu, ed. (1991). *Kasomondai to chihojichitai,* 15. Taga Publishers, Tokyo (in Japanese).

National Land Agency (1996). *The Present Condition of Depopulation Countermeasures, 1995 ed.* 361 pp. (in Japanese).

Report of the Local Sectional Meeting of the Economic Council (1967). *Koomitsudo keizaishakai eno chiikikadai,* 2 (in Japanese).

16

New Relationship between Forests and City Dwellers in Japan

Yoshiya Iwai

During the past several decades, forestry production in Japan has undergone a drastic decline. The volume of log production has fallen from a peak of 52 million m³ in 1967 to 18.7 million m³ in 1999. Annual reforestation area has decreased by one-eighth, from 430,000 ha in 1954 to 33,860 ha in 1999.

What changes, then, have rural communities supported by forestry gone through during this period? The main industries in rural areas in Japan have been agriculture and forestry. Rice and other crops have been cultivated on small and low-yield farmland, while timber has been harvested from forests.

In the 1960s, people in rural areas began to move to cities, partly because agriculture and forestry were not profitable. Note that this population outflow occurred when Japan, with its high economic growth, was rapidly joining the other developed countries of the world. The population outflow from the country to cities has continued to the present, and rural depopulation has created various problems. Today young people are rarely found living in rural villages. In particular, small villages of less than 100 families have difficulty carrying out community activities such as repair to footpaths, water management, and removal of snow. Many such communities have disappeared, and in many of the remaining ones farmland is no longer cultivated and man-made forests have been abandoned. Weeding and thinning are not performed.

What are the effects of these changes in Japan, a country located in the temperate monsoon climatic zone, where rainfall is concentrated in the summer rainy season? Since Japan is a mountainous country with very steep slopes, concentrated rainfall often causes landslides, erosion, and floods during the rainy season. In the dry season, on the other hand, an unstable water supply from rivers becomes a problem. Abandoned rice fields lose their capacity to hold water. Neglect of man-made forests also results in poor understory vegetation, which in turn degrades the forest's water-holding function as well as its erosion-control function. In this regard, poor management of rice fields and man-made forests is a serious concern with regard to water management in Japan.

With respect to forest resources, abandoned man-made forests are more vulnerable to wind and snow damage because trees are thin and weak. If trees are no longer regarded as timber resources for the future, owners are further discouraged from managing them, leading to a further degradation of the forest resource. The decline in forestry production and degradation of forest resources in Japan are thus two sides of the same coin. They form a vicious circle, resulting from depopulation in rural areas and the decline of the forestry industry.

In Japan forest management and forestry production are performed by people in rural areas, whereas to city dwellers forests have mainly become recreational sites for mountain climbing, hiking, and camping. City dwellers see the forest only while walking or looking through car windows. In the past two decades, however, the relationship between forests and city dwellers has changed dramatically. People in urban areas have begun to pay more attention to forests and have become interested in working in them. Some examples are given in the next section.

Different Approaches to Forests by City Dwellers

Case 1: From "Investing" to "Participating" in Reforestation

In Japan profit-sharing reforestation has been widespread in both private and public forests, including national forests. In Shiga Prefecture, a profit-sharing reforestation system was introduced in 1983 for private and public forests. One contract unit is about 0.1 ha of a 12- to 30-year-old man-made forest for ¥300,000, and investment from city residents is encouraged. With the money invested, forest owners take care of and manage their forests by performing weeding, thinning, etc. The forests are clear-cut and sold at the age of 55. Profits are shared equally by forest owners and investors. In Shiga Prefecture, contracts for profit-sharing reforestation were drawn up for 56 ha of forests, and 533 people invested until 1995. The primary concern of investors lies in contributing to forest resource development, rather than in associated profit making.

An unexpected outcome is that investors in Shiga have become interested not only in investment but also in working in the forest. Since profit-sharing reforestation contracts prohibit investors from working in the particular forest, the investors contacted other forest owners and began weeding in their forests a couple of years ago. Under supervision of the owner, they now engage in weeding and straightening trees bent by snow three times a year.

A profit-sharing reforestation system was introduced in the national forests by the Forestry Agency in 1984; approximately 25,000 ha were allotted for this purpose as of 1998. More than 85,000 investors are involved, 8,400 of whom have set up a friendship society that provides its members with opportunities to experience forestry work and a variety of other activities in order to better understand forests.

Case 2: Profit-Sharing Reforestation by Companies in National Forests
The Forestry Agency (in 1992) has initiated a new type of profit-sharing forest scheme that targets companies as investors with the objective of developing healthy forests as well as promoting a wider understanding of forests among the general public. (Until then, there was no profit-sharing reforestation scheme involving companies within the national forests.) Each block is 1-5 ha, and companies may plant conifers and/or broad-leaved trees and take care of the forest for a maximum of 80 years. The companies are responsible for the cost and labour necessary for planting and nurturing the forest. At the time of cutting, 70% of the profit goes to the companies and the remaining 30% to the Forestry Agency.

From 1992 to 1995, profit-sharing reforestation of this type was carried out on a total of 85 blocks, or about 223 ha. Firms with a strong interest in the environment and high expectations of a good public relations spin-off have participated in this scheme. Nippon Life Insurance Company, in particular, has carried out a "1 million tree planting campaign" as a philanthropic activity, and 500,000 trees were planted by the end of 1997. It is noteworthy that in such a project involving company partnership, labour is provided by ordinary citizens and employees who have previously had little direct contact with forests.

Case 3: Volunteer Activity by Students
In 1974 an anti-herbicide campaign was launched to protest a plan to apply herbicide in a 300 ha forest in Toyama Prefecture. A university professor who organized the campaign not only protested against the plan but also proposed an alternative and put it into action. He appealed to university students in Tokyo to participate in volunteer weeding (to make herbicide unnecessary). Eventually Toyama Prefecture started supporting this activity with a subsidy, and since 1985 (10 years later) a daily allowance of ¥4,000 has been provided to the volunteer students, although they are still responsible for their transportation to Toyama. During the first 12 years, a total of 1,600 students volunteered and weeded 1,200 ha of the forest. This activity has been carried out annually, and by last year a total of 2,800 students had volunteered. Their motivation ranges widely from self-discipline to nature protection to earning money (CCYAJI 1986, 102-103).

Another example of volunteer forestry was work carried out by a student group from Waseda University, in Tokyo. It was initiated as volunteer work to plant trees after a large fire in 1967 in Iwate Prefecture (northeastern Japan). Since then, the students have visited the area every year, performing forestry and farming work with the local people. Besides the work, interaction with the local people has been an enjoyable and valuable experience for the students. The group also developed an international project with Earlham College in Indiana. Students from both schools work together in farms and forests in a rural village in Iwate every summer. With the experience gained from this project, one of the students from the group has assumed leadership in a local community environmental group in Tokyo (CCYAJI 1986, 122-23). From 1974 to 1996, the total participation amounted to 26,196 worker-days, and 1,611 ha of forest was weeded.

Case 4: Volunteer Activities by Citizens

Shinrin Club

In 1983 a group called Shinrin Club (Forest Club), which had engaged in volunteer work for the previous eight years at farms and in planting trees at sites belonging to national hospitals, was reorganized. The group made a fresh start with the purpose of "pursuing an ideal relationship between forests and people as well as efficient utilization of trees by human society." They are interested in forestry, which, in their opinion, means not only the production of usable wood but also the maintenance of forests.

This group entered into their first contract with the Forestry Agency for profit-sharing reforestation in 1984, and began clearing, planting, and weeding in a 1.6 ha forest. This was followed by a second contract for a 1.1 ha forest in 1986. Today the group has a hundred members. Their main activity is normal forestry work in profit-sharing reforestation areas, occasionally with additional recreational activities, such as *sansai* (forest vegetables) cooking, moonlight parties, and hiking. Another activity is running a summer school in the forest for children, in which the children stay in a forested area for five to six days and learn about the forest and various ways of playing in a natural setting (Shinrin Club 1994).

Somanokai

Somanokai (Group of Forest Workers) was set up by city residents as a forest owners' group in the early 1980s. Mr. "I" had been interested in nature and wanted to use firewood, so he searched for a forest where he could collect wood. He finally found one in Shiga Prefecture, but it was 25 ha, too large to manage by himself. He asked other people, including nature lovers, to become part owners in the forest and they agreed to do so. The joint-owner families visit the forest every weekend and participate in performing general maintenance of the coppice forest, making charcoal, chopping firewood, and gathering forest vegetables – all healthy activities. The members share the idea that they should review their "modern" lifestyle by learning from lifestyles in rural areas, where, in the past, people made a living from forest resources by producing firewood and other wood products. Through activities in the forest, they learn a great deal from nature in order to reflect on their present life. They prefer natural broad-leaved forests to man-made forests. The group had 108 members as of 1997.

Hamanokai
Hamanokai (Group of the Seashore) was organized in 1986 by a group of people who had helped restore forests in the Kanto area that had been heavily damaged by snow that year. After consultation with forest owners in rural areas in near Tokyo, the members go to forests that are no longer maintained, and perform weeding and thinning on the first Sunday of every month. What is characteristic about this group is that they view the activity as "leisure forestry" and enjoy it as a sport rather than as a means to contribute to forest management. Some members have also set up another similar group. The group had 150 members in 1997.

Tokyo No Ringyoka To Kataru Kai (Let's Talk to Forest Owners in Tokyo)
Tokyo No Ringyoka To Kataru Kai, established in 1993, holds discussions with forest owners in Tokyo about forests and forestry. They also arrange "women's weeding tours." The aim of such activities is to understand the reality of forestry in Tokyo and to consider the future of Japanese forestry.

In February 1996 there was a conference of citizen groups involved in forestry activities such as those described here. A panel discussion entitled "Creation of Forests by Citizens" and a number of meetings were held, in which 850 people participated at their own expense. The discussions covered a wide range of issues such as a nationwide networking of citizen groups and the importance of the role of administrators. It was pointed out that forest owners need to accept responsibility for forest management, and citizens also need to recognize the importance of forest issues. It is remarkable that forestry activities engaged in by the general public have developed to such an extent throughout the country.

Case 5: Forest Experience Programs/Courses Sponsored by Companies and Local Governments
Asahi Shinbun, one of the largest national newspaper companies in Japan, launched a "Green Campaign" in 1978 to commemorate its centennial anniversary. It acquired a 148 ha forest in Shiga Prefecture, and in the

following year founded a facility for research and training there. Various courses have been offered to citizens, including courses on weeding, outdoor activities, and forest vegetable cooking. Seminars and symposia relating to forests and trees have also been held. More than 10,000 people a year participate in these activities and experience a direct relationship with forests.

Local prefectural governments, cities, and villages have also offered courses on forests/forestry experience quite frequently in recent years. Most programs are usually one-day programs for 50-60 participants, but some offer weeklong programs on forest maintenance activities such as weeding. Active participation by young women is conspicuous in these courses.

Case 6: Forests for Lease

In 1994 a business renting out forests was initiated in Gunma Prefecture. A 5 ha forest under common ownership was divided into 50 0.1 ha blocks and leased to people in urban areas for 10 years. Lessees pay an annual rent of ¥5,000-15,000 yen (depending on the condition of the forest) and carry out forestry work such as weeding and pruning on their blocks. They are permitted to build a hut or shed on the property. This may be described as a forestry version of *Klein Garten* (originally from Germany).

Local government officers provide instruction on how to maintain the forest. The necessary equipment is also made available free of charge from the local government office. In the first two months, contracts were signed for 19 blocks. The lessees live in Tokyo or surrounding areas, and it takes some of them five hours to get to the forest. Their primary objective is to grow trees and enjoy working in the forest.

A similar business was started in Gifu Prefecture. The idea was that forests, which are poorly managed due to a shortage of forestry labour, are leased out to people from the city so that they can participate in growing trees and mushrooms in fresh forest air, and the forest, in turn, is well maintained. The contract extends over a period of 10 years and the admission fee is ¥100,000. Lessees choose between two block sizes, 0.025 ha and 0.05 ha, with annual rents of ¥11,250 and ¥22,500, respectively. The rental prices were based on the average rent for land used for golf courses. Lessees hold ownership of standing trees for 10 years on the condition that one-third of trees should remain standing at all times. Most lessees are in their 20s or 30s, and work for companies or are self-employed. Some express a wish to buy the land in the future so that they can build a log house. Because this business has been so popular, there are plans to offer another 1.5 ha of forest for lease.

Case 7: Employment by a Working Team from a Forest Owners' Cooperative

With funds from local forest owners, forest owners' cooperatives in Japan carry out silvicultural operations and log production for the owners. In recent years, it has become extremely difficult for forest owners to continue silvicultural management on their own, making the role of the cooperatives more important. Unlike in the past, today forestry labour cannot be adequately provided by the local population because most young people leave rural communities to live in urban areas. Consequently, several forest owners' cooperatives began changing their philosophy in the 1990s and tried to attract workers from cities. They placed advertisements in classified ad magazines popular with young city people. To their surprise, they received many applications.

One such case is a cooperative in Nagano Prefecture, where the Winter Olympics were held in 1998. This cooperative attempted to recruit 10 forestry workers through a classified ad magazine and received 44 applications in 1993. It interviewed the applicants in detail, organized a field (orientation) trip around the village in order to give them a picture of the current situation, and finally decided to employ 12 applicants, ranging from 21 to 51 years old, including three women. All had worked in cities until then and had decided to change occupation to work for the cooperative. Since there is no vocational training system for forestry workers in Japan (unlike in European schemes), the successful applicants needed to learn the techniques under the supervision of skilled and experienced workers for at least two years.

What motivated the new employees to work in the cooperative? Mr. "A," who is 28 years old and who had taught at a high school, said, "I have always wanted to live near forests or the sea. Before I started working in the forest, I was never as grateful for the sun as I am now." Mr. "B," who is 23 years old and whose wife also obtained a job in the same cooperative, said, "I can enjoy both rural life and my hobby, mountain climbing, at the same time." Mr. "C," who is 21 years old and was a machinery designer, said, "I wanted a job that required physical effort. I was also interested in growing trees."

Nowadays, many forest owners' cooperatives throughout the country recruit workers from urban areas, and it is no longer uncommon for a cooperative to have several workers who are from cities but who have changed jobs to work in forests. Although not an example of a forest owners' cooperative, a group of former members of Hamanokai (mentioned earlier) also entered into contracts with forest owners to perform forestry work such as weeding on weekends and holidays, thereby earning income. These members have their own regular jobs but make additional income from forestry work.

Case 8: Forestry Work as Therapy and Education for Mentally Handicapped Children

In 1974 Mr. "T," a forest owner in Shiga Prefecture, began holding summer camps for mentally handicapped children as a volunteer activity. From the experience he gained, he also developed the idea of making the children do forestry work every weekday after school. A daily regime was begun where the children participated in growing shiitake mushrooms and planting trees in the forest. A couple of months later, it was observed that epileptic children had no fits during the work, and those who suffered from autism started to smile and talk to others. Although these positive changes cannot be directly attributed to the forestry work, Mr. "T" believes in the power of the forest to heal mental illness and has continued the program (CCYAJI 1986, 206-207).

Motivations of People Who Want to Relate to Forests

The preceding examples of eight types of forestry activity show only superficial fragments of the participants' motivations and the group leaders' ideas. No comprehensive research or analysis has been conducted concerning the motivation of the participants. This section will therefore examine in detail the motivation of city people to participate in forestry volunteer work, based on a survey conducted in Yamashiro Town.

Yamashiro Town is in Kyoto Prefecture, located 30 km from Kyoto and Osaka. It was formerly a rural area but its population has increased recently because its location is convenient for commuting to the cities. Yamashiro Town owns a forest park that had degenerated in recent years due to poor management. In 1996 the town drew up a plan to attract volunteers to maintain the forest. It conducted a survey on the occasion of recruitment of forestry work volunteers.

Applicants were expected to be 16 years or older, to have access to the forest, and to participate in the activity more than seven times a year. The advertisement read: "Why don't you join in the volunteer activities of our program and get involved in the development of a healthy forest with other participants from inside and outside Yamashiro Town? You will enjoy a variety of activities in the forest such as growing Matsutake mushrooms, raising kunugi (*Quercus acutissima*) trees and constructing a wild bird sanctuary. Other activities include pruning, thinning, building log houses, and making charcoal." In the advertisement, the forest was described as a combination of a 40-year-old man-made forest (*Chamaecyparis obtusa* and others) and a natural forest (*Quercus serrata* and *Querucus acutissima*, etc.).

The town wanted 20 volunteers but worried about whether there would be enough applicants. Despite this concern, 290 applications were received, thanks partly to a major newspaper's article highlighting the program. The town office was faced with the unexpected difficulty of selecting 20 successful applicants, and finally decided to accept all of them.

Let us now look at the characteristics of the applicants.

Table 16.1

Age and sex of applicants for forestry volunteer work in Yamashiro Town, Kyoto Prefecture, January 1996.

Age group	Male	Female	Total	%
< 20	6	4	10	3.4
20-29	33	33	66	22.7
30-39	34	9	43	14.8
40-49	63	12	75	25.8
50-59	61	10	71	24.4
60-69	19	1	20	6.8
³ 70	2	0	2	0.6
Unknown	3	0	3	1.0
Total	**221**	**69**	**290**	**100.0**

Source: Kyoto Prefectural Government (1996), unpublished document.

Age and Sex

Table 16.1 shows the age and sex of the applicants. It is worth emphasizing that 25% were women. Of those in their 20s, 50% were women, indicating that women at this age are particularly interested in volunteer forest work. Regarding age, 25% of applicants in their 40s represented the largest segment of the surveyed population, and those in their 50s the second largest. Applicants in their 30s accounted for 14.8%.

Residence

Table 16.2 shows the distribution of applicants by place of residence. People from local areas, such as Yamashiro Town itself and the Souraku County (where Yamashiro Town is located), were in the minority. The largest group, 45.8%, lived outside Kyoto Prefecture; most were from large cities such as Osaka City and Nara City. An overwhelming majority, including those from Kyoto City, lived in big cities and urban areas.

Table 16.2

Residential areas of applicants for forestry volunteer work in Yamashiro Town, Kyoto Prefecture, January 1996.

Residential Area	%
Kyoto City	20.0
Yamashiro Town	5.8
Souraku County (except Yamashiro Town)	8.9
Kyoto Prefecture (except Kyoto City and Souraku County)	19.3
Outside Kyoto Prefecture	45.8
Total	**100.0**

Source: Kyoto Prefectural Government (1996), unpublished document.

Table 16.3

Occupations of applicants for forestry volunteer work in Yamashiro Town, Kyoto Prefecture, January 1996.

Occupation	%
Company employees	38.9
Students	11.3
Public servants	9.3
Teachers	6.8
Housewives	6.2
Unemployed	3.7
Agriculture and forestry	0.3
Others	23.5
Total	100.0

Source: Kyoto Prefectural Government (1996), unpublished document.

Occupation

As shown in Table 16.3, company employees represented the largest group, 38.9%, of the surveyed population, followed by students (11.3%), public servants (9.3%), and teachers (6.8%). It is striking that only 0.3% of applicants had had jobs in agriculture and/or forestry.

Motivation

Table 16.4 outlines applicants' motivations for applying. The applicants indicated their motivation in the survey. Some participants indicated more than one motivation, leading to a total of 376 answers. Table 16.4 shows only the 346 answers that were the responses given by more than one applicant.

The answers fell into four categories:
- *Type I: Contribution to society.* "To understand the importance of forests" was mentioned by 40 applicants and "to learn about forestry" by 19 applicants. In both answers, applicants expressed their intention to learn more about the global problem of forest resources as well as forestry. "To protect forests and contribute to society" indicates that the applicants wanted not only to understand the situation of forests and forestry but also to be active. In the Type I category, there were 118 responses, constituting one-third of total answers.
- *Type II: Nature/hobby-oriented.* There were 53 answers of the type "to work in a natural setting and to experience forestry work." Of these, several answers were as follows: "I have seen forests only on occasions of hiking, mountain climbing and driving, but now I want to have a closer

Table 16.4

Motivations of applicants for forestry volunteer work in Yamashiro Town, Kyoto Prefecture, January 1996.

Type	Motivation to apply	Number of responses
I: Contribution to society	To understand the importance of forests	40
	To learn about forestry	19
	To protect forests and contribute to society	59
	Total	118
II: Nature/hobby-oriented	To work in a natural setting and to experience forestry work	53
	To experience charcoal making, log cabin building, and mushroom growing	42
	To enjoy the outdoors	7
	To enjoy nature and forests	20
	To have a direct relationship with nature	27
	To learn about trees and forest ecosystems	11
	Total	160
III: Step towards living in a natural setting	To live in a natural setting in the future	13
	To get a job in agriculture and/or forestry	11
	To get a job related to nature and forest	7
	Total	31
IV: Others	To teach children about nature and forests	13
	To meet people	12
	To be thankful for nature and forests	2
	For self-discipline	2
	To supplement tertiary education with field experiences	2
	Background in rural areas	4
	To learn how to deal with nature	2
	Total	37
	Total responses	346

Source: Kyoto Prefectural Government (1996), unpublished document.

relationship with the forest by working in it." There were 42 answers of the type "to experience charcoal making, log cabin building, and mushroom growing," 7 applicants wrote "to enjoy the outdoors," and 20 wrote "to enjoy nature and the forest." "To learn about trees and forest ecosystem" was mentioned by 11 applicants, indicating their attachment to nature and their desire to experience it at a closer distance. Twenty-seven applicants indicated, "to have a direct relationship with nature," and some of them expressed their desire "to escape even temporarily from a life in a small apartment or a life in the hectic city." All these answers indicate the applicants' attachment to nature as well as their desire to reduce daily stress.

- *Type III: Step towards living in a natural setting.* Thirteen answers of the type "to live in a natural setting in the future," 11 answers expressing the desire "to get a job in agriculture and/or forestry," and 7 answers indicating a wish "to get a job related to nature and forests" were recorded. These people applied for volunteer work as a step towards their future life and jobs in natural settings.

- *Type IV: Others.* Type IV includes other motivations that do not fit into the above three categories. Thirteen applicants said that they wanted "to teach children about nature and forests," meaning education either at home, at school, or both. They want forests to be a place of education for future generations. "To meet people" was mentioned by 12 applicants, referring to meeting people from rural communities and/or volunteers from other areas. This indicates the possibility of extending human relationships through forestry volunteer activities. "To be thankful for nature and forest" was found in 2 answers, one of which explained as follows: "I want to express my gratitude for the lost forest which had existed on the land where my house is now." The other answer was, "I want to express thanks for the forest which my children always play in." Other answers – "for self-discipline" and "to supplement tertiary education with field experiences" – were given by 2 people. The latter answer can be summarized as follows: "I have majored in forestry/environment at a university in a big city, but I could not obtain appropriate field education." This answer should make one reconsider the forestry/environment education provided at universities. Although these answers cover a range of motivations, many concern education in the broader sense, for their children or for themselves.

Conclusion

Having looked at different approaches to forests by urban dwellers, and the motivations of these urban dwellers, we turn now to the social backgrounds behind the trends and their significance.

As already noted, Japan experienced population inflow into cities and dramatic economic development in the 1960s. As a result, urban areas suffered from air and water pollution caused by exhaust and drainage from factories. People in cities then became aware of environmental problems, although forests were not part of this perspective at the time.

Many cities in Japan are located midstream or downstream on rivers, and depend on the rivers for their water supply. Water from rivers is often in short supply in the dry season, however, partly because of increased demand for water in cities. This situation gradually made city residents aware that their own city life was influenced greatly by forests upstream. In addition, the critical situation of tropical rain forests has been highlighted by the mass media. Japan is the largest importer of tropical timber from Southeast Asia. Realizing the huge volume of tropical timber being used for their own houses and furniture, many Japanese have become aware of the crisis of tropical rain forests as not so much a distant, alien problem but one directly involving Japan. People have also learned that many non-governmental organizations (NGOs) throughout the world have been making efforts to protect and restore tropical rain forests in Africa, the Amazon, and Southeast Asia. This has made them question their own attitudes of only thinking rather than doing something.

In this way, many Japanese, especially those living in cities, have come to appreciate the importance of forests and nature. They now want to have a deeper understanding of forests and to personally contribute to management and protection by concrete action. This is reflected in the 34% of responses to the Yamashiro Town survey that were classified as Type I responses ("Contribution to society"), and the first five approaches described in the section "Different Approaches to Forests by City Dwellers."

The economic development that led to deterioration of the environment in cities improved the living standards of its residents at the same time, so that they achieved almost the same standard working hours as in other developed countries. People are now interested in how to spend their free time meaningfully. This is reflected in the overseas travel boom and in the popularity of outdoor activities, especially those having to do with forests.

About 20 years ago, there was a second-house boom among famous television personalities in Tokyo, who purchased farms with some forest within 200 km of Tokyo for weekends. They wanted physical exercise in the forest, such as chopping firewood, repairing log houses, and generally maintaining the forest. The same motivation is basically reflected in Type II, "Nature/hobby-oriented." An example of this type is the "forest for lease" approach. Hobby-oriented motivation can also be observed in cases 3, 4, and 5 in the same section ("Different Approaches to Forests by City Dwellers"). Some people, unsatisfied with "a hobby," wanted to gain employment in forests and live in rural areas. This motivation is categorized as Type III and illustrated by case 7.

As shown in the examples above, city dwellers have become more intimate with forests. In order to realize their goals, they needed access to forests in rural areas. Until that time, forest owners in rural areas, whether private, public, or national, had never expected that people from cities would come and work in forests. Even now, some forest owners do not view it seriously, believing that it is impossible for city people to perform such work as weeding, thinning, and pruning.

In rural communities, however, forest management and silvicultural operations have been abandoned in many forests because of depopulation and the declining profitability of forestry. Rural communities are in such a depressed state that people there can no longer say, "It is impossible for city people to carry out forestry work." Now rural communities expect city people to help them with forestry work, to utilize their forests, thereby earning more income for the local community, and to better understand forests and forestry work. They have therefore gradually made their forests available to city people. This could never have happened if the rural communities had been thriving and the forestry industry had been sufficiently profitable. The crisis in both Japanese rural communities and forestry encouraged rural communities to accept people from urban areas.

Finally, reference should be made to the significance of the fact that city dwellers are relating more closely to forests. As previously explained, people in cities have recognized the meaning and benefit of working in forests. They learn from the forest and find new values there while actually working in the forest. One of the things they learn from such an experience is that the forest is an ecosystem consisting of different elements, such as flora and fauna, soil, and water. They also learn how nature works and how beneficial the forest is to human society. It makes them understand how human beings should deal with forests in order to continue benefiting from them.

Recognition of the values of forests by city dwellers will, in turn, encourage people in rural areas, including forest owners, to review forest management and forestry. The point is that both city and rural dwellers can cooperate together, and not confront each other as they often did in the past (for example, over the question of whether forest preservation or forestry should be the priority). If they can cooperate on issues concerning forests and forestry, people in rural areas will feel comfortable with, and be reassured by, such a partnership.

As the tendency of city dwellers to become closer to forests grows, they will voice concern over global problems of forests and forestry. The case of the university in Tokyo and a college in the US cooperating together in forestry work is a good example of opportunities for international partnership for global environmental protection. It is hoped that a newly evolving relationship between forests and city people will produce a new perspective on forests and forestry that may eventually give birth to a new forest culture.

References

CCYAJI (Central Commission of Youth Action on Japan Island) (1986). *Ganbare Mizu to Midori*, 70-247. CCYAJI, Tokyo (in Japanese).

Kokudo Ryokka Suishin Kiko (1998). *Shinrin Volunteer no Kaze*. 241 pp. Nippon Ringyo Chosakai, Tokyo (in Japanese).

– (1995). *Juminsanka ni yoru Ryuiki no Morizukuri*. 144 pp. Kokudo Ryokka Suishin Kiko, Tokyo (in Japanese).

Shinrin Club, ed. (1994). *Watashitachi no Morizukuri*. 170 pp. Shinzansha Shuppan, Tokyo (in Japanese).

17
Treatment of Forests and Wildlife in Modern Society
Atsushi Takayanagi

At present, the Forestry Agency (FA), which is in charge of forest and forestry policy in Japan, is not directly concerned with wildlife policy. In 1971 the Environment Agency (EA) was established and assumed responsibility for wildlife from the Forestry Agency. The FA, however, significantly influences wildlife, since forests, which cover more than 60% of Japan, are the most important habitats for many forms of wildlife. People are strongly interested in wildlife, and this cannot be ignored by the FA when they deal with forest policy. The relationships between forests, humans, and wildlife will be discussed against this background.

Development of Hunting Laws and Wildlife Protection Movements in Modern Japan

The Meiji restoration in 1868 ended the long-lasting feudal system in Japan and prompted the reconstruction of the social system.

Hunting had been one of the privileges of rulers from ancient times, and they would enclose land for hawk hunting. In the highly feudal Edo era (1603-1867), ordinary people were prohibited from hunting, with the exception of a few professional hunters (Forestry Agency 1969a). Although farmers could not own weapons, there were usually many guns in villages, used for protecting farms from wild animals by frightening them away with false fire (Tsukamoto 1983). Commoners who shot animals without permission were strictly punished for poaching (Endo 1994). The fact that the government permitted farmers to have guns suggests that there may have been an abundance of wildlife, and that damage caused by such wildlife was so severe that farmers could not live without a means of defence (Sakurai 1995).

In the Meiji era, the ban on hunting by commoners was lifted and people began to shoot animals freely regardless of place and season. The government quickly drew public attention to the danger of such practices in 1868, the first year of the Meiji era, but exploitation of wildlife and shooting accidents still occurred in both cities and the countryside (Forestry Agency 1969b, 6-30). The government began to take control of hunting in suburban areas and prohibited the shooting of birds in previously protected areas the next year. One year after this, firing of guns in residential areas was prohibited.

The Gun Control Regulation was introduced in 1872, bringing with it a licence system for gun trade and hunting with guns. This regulation was the predecessor of the current Gun and Gunpowder Act. In the next year, the Regulation on Birds and Animal Hunting was published. It included registration of hunting licences, hunting tax, and age limitation for licences, and placed limitations on the types of gun, hunting season, forbidden deeds, and places available for hunting. These regulations formed the framework of the subsequent Game Laws. Even after publication of these laws, it was very difficult to control hunting, and revisions were required every year.

The appearance of the Murata gun, increases in gun imports, and the growing demand for stuffed animals and feathers increased poaching and exploitation of wildlife, resulting in clear decreases in the numbers of wild animals and birds. As it was necessary to protect wildlife, the Hunting Regulation was published in 1892 to regulate hunting more strictly. This regulation separated licences for gun hunting from those for other styles of hunting, designated protected species, and established seasons during which hunting of several species was prohibited. These new rules, however, were intended to protect only wildlife beneficial to agriculture and forestry; other species were to be exterminated as vermin.

The first Hunting Law (Law No. 20) was published in 1895 and was little different from the Hunting Regulation; the distinction between professional and amateur licences was abolished, collection of eggs and young of protected birds was prohibited, and common hunting areas were approved. These common hunting areas were former conventional lands or ponds, which were maintained by a local community. The members of the community cooperated in hunting in these areas, and the earnings were allotted to public operations or allocated to each family.

The Hunting Law has been revised frequently, and one of the largest changes (Law No. 32) was made in 1918 when game species were designated instead of protected species. In reality, this did not have much effect on hunters because all mammalian game species with the exception of Amami rabbit (*Pentalagus furnessi*) were designated as game. However, the basic attitude towards wildlife was reversed from one in which the majority was to be hunted to one in which most species should be protected. No new common hunting areas were allowed, and hunting districts managed by governmental organizations were established.

Just after the Second World War, Dr. Oliver Austin of the US Department of Natural Resources recommended promotion of wildlife protection in

Japan. During the American occupation, the Japanese government revised the Hunting Law. In 1947 the list of game birds was halved, and otter, wildcat, monkey, and female deer were taken off the list of game species. It became illegal to hunt protected species without permission, and wildlife reserves were established.

The next major revision was in 1963. The law was renamed the Wildlife Protection and Hunting Law, and a new purpose was stated: to improve living circumstances of man and promote agriculture and forestry through wildlife protection by controlling hunting. The main changes from the Hunting Law were establishment of the Wildlife Protection Action Plan. The plan allocated areas in which hunting was temporarily prohibited, and introduced the prefecture hunting licence system, in which the license was valid only in the prefecture where it was issued.

One year after the EA became the authority for wildlife administration, the Natural Environment Conservation Law was published. In 1978, the EA revised this law according to a report submitted by the Natural Environment Conservation Council concerning the proper administration of wildlife protection and hunting. The EA fixed the valid period on licences, abolished the prefecture hunting licence system, and introduced a registration system in which hunters had to register themselves in each prefecture where they wished to hunt. The establishment of private hunting areas was also permitted under the revised law.

In 1994 the game species list was revised. Japanese squirrel (*Sciurus lis*) and Japanese giant flying squirrel (*Petaurista leucogenys*) were taken off the list, and exotic animals – masked palm civet (*Paguma larvata*) and common raccoon (*Procyon lotor*) – were placed on it (Environment Agency 1994). Deer, including females, was listed again since deer numbers had increased and these animals had caused severe damage to crops and plantations in several regions.

The law was extensively amended in 1999. Prefectural governments were placed in charge of giving permission to capture wildlife. Before the amendment, the national government had had the authority to grant this permission, and delegated it to prefectures. A new management system, the Specified Wildlife Conservation and Management Plan, was established. When a species show extreme increase or decrease in the local area, the prefectural government can make a Specified Wildlife Conservation and Management Plan and manage the species according to the plan. Under the plan, the prefecture can ease hunting restrictions determined by the EA.

Social Status of Hunting, and Its Control

Wildlife management consists of protection, hunting, and damage control (Takayanagi 1991). In this section, problems concerning hunting will be discussed.

In modern societies, hunting is generally regarded as a hobby because wildlife meat, leather, and fur are not necessary for our daily life. Few people obtain any income from hunting, and there are no professional hunters for whom hunting is the sole source of income. Despite its status as a hobby, hunting also has many social implications. For example, hunting affects animal numbers. Reducing animal numbers impacts the ecosystem but may result in less damage to crops. Without protection, agriculture and forestry are impossible in mountainous areas since crop fields are adjacent to wildlife habitats. It is therefore necessary for rural communities to reduce wildlife numbers by hunting and/or culling. Some restaurants serve venison or other wildlife meat, and shops in mountain villages deal in goods made of fur or antlers. These materials can be supplied by hunters. Animals in the wild are a valuable natural resource for animal and bird watchers and photographers. Although hunting is a hobby, it has social consequences.

Hunting must be discussed from various points of view. The motivation for hunting is one of the most important issues in the modern world. In developed countries, there is no need to hunt wildlife for food or because of some ill-defined "hunting instinct." Some people hunt wildlife to sell their meat or skin, to present them to friends, to protect crops, and so on. Motivation suitable for the society promotes proper hunting. Recently the average age of hunters has gone beyond 50, because the younger generation generally has no motivation to hunt. In Japan the police strictly control gun ownership, and hunting is not particularly popular with urban dwellers. However, a strong economic motivation for hunting would result in poaching, which may destroy local wildlife populations. Problems concerning hunting cannot be solved by laws and regulations alone. People must examine their own motivations for hunting.

The same difficulties are encountered in management of wildlife as a forest resource. The cost of damage caused by large mammals is too high to be compensated by income from wildlife products (Takayanagi 1993). Hunters do not want to maintain such game with the price being payment of farmers for the damage caused by these animals. In the early 20th century, common hunting areas were kept to obtain income based on the availability of wildlife, mostly waterfowl or other bird species (Table 17.1). These species are seldom responsible for economically significant damage.

Hunting must be controlled in a local area in order to manage wildlife as a local resource. The designation of hunting districts is a typical and popular system for this purpose. In modern Japan, however, there is little motivation to manage wildlife as a local resource. There are about 40 hunting districts in Japan. Only a few are capable of generating large income, while the rest have no economic importance (Takayanagi 1995), existing because of local hunting traditions and/or enclosure. It is very difficult to maintain a good balance between damage and profit, especially in the case of large

Table 17.1

Common hunting areas (CHA) in the early Showa era, 1926.

Prefecture	CHA	Landscape	Area (ha)	Licence holders	Area/holder (ha)	Reduction of profit to licence holder	Allowance to other hunters	Game species and average bag per year
Iwate	Toen	Forest	1,410.5	16	88.2	Distribution of profit	With charge	440 pheasants, 1 deer, 1 bear
Ibaraki	Kiriike	Irrigation pond	6.4	36	0.2	Tender	Pay by game	2,500-3,500 ducks
	Kaminoike	Pond	435.7	174	2.5	Hunter's income	Not allowed	300 ducks
Chiba	Fusei	Marsh	342.4	49	7.0	–	Not allowed	4,000-7,500 ducks
	Sibazaki	Marsh	151.4	84	1.8	Distribution of profit	–	400-600 ducks
	Hananoi	Marsh	153.7	133	1.2	Tender	–	500-800 ducks
	Omuro	Marsh	205.0	117	1.8	Tender	–	200-500 ducks
	Teganuma	Marsh	921.7	349	2.6	Allowance	–	6,000-7,000 ducks
	Konosaki	Riverside	59.2	158	0.4	–	–	ducks
	Otake	Irrigation pond	27.1	28	1.0	–	–	4,000-5,500 ducks
	Naganuma	Swamp	375.6	103	3.6	–	–	300-500 ducks
	Onuki	Riverside	12.4	4	3.1	–	–	1,000 ducks
	Yujaike	Irrigation pond	21.0	18	1.2	–	–	500-1000 ducks
Nigata	Oike	Pond/hill	97.1	6	16.2	–	–	500 ducks
Fukui	Kokubu	Crop field	158.7	–	–	–	–	ducks
Toyama	Minokuma	Irrigation pond	75.8	9	8.4	–	–	1,000-4,000 ducks
Shizuoka	Jittari	Forest/natural field	448.0	6	74.7	–	With charge	250 pheasants, 110 rabbits, 15 wild boars

Aichi	Fujie	Irrigation pond/ rice field/forest	9.7	2	4.9	Tender	With charge	1,500–3,000 ducks
	Higasisakai	Pond	121.1	4	30.3	–	–	1,000–1,500 ducks
Osaka	Yamada	Pond	523.2	–	–	Tender	–	1,500–3,500 ducks
	Muroike	Pond	87.4	115	0.8	Tender	–	1,500 ducks
	Shakaike	Pond	141.9	–	–	Fee	–	4,000 ducks
	Hachoike	Pond	128.0	–	–	Tender	–	4,000 ducks
Ehime	Nisigaikai	Island	4.5	6	0.8	–	No charge	4 deer, 1 monkey
Fukuoka	Yokowai	Hilly country	80.3	4	20.1	–	–	1,000–2,500 ducks
	Misawa	Hilly country	154.2	2	77.1	–	–	1,000–3,000 ducks
	Nisijima	Hilly country	65.4	2	32.7	–	–	1,500–3,000 ducks

Note: Dashes signify unknown data.
Source: Forestry Agency (1969), *History of Wildlife Administration*, 186–92.

mammals. Assuming that damage problems can be resolved, the future of hunting depends on whether local people can be motivated to maintain wildlife populations as resources.

Damage Caused by Serow and Deer

Damage problems indicate most clearly the relationships between humans and wildlife in societies such as Japan where people are hesitant to discuss hunting openly. The main pest mammals cause a variety of types of damage: voles (mainly *Clethrionomys rufocanus*, *Eothenomys smithii*, and *Microtus montebelli*; barking), hares (*Lepus timidus* and *L. brachyurus*; barking and leaf eating), sika deer (*Cervus nippon*; barking, leaf eating, crop eating), Japanese serow (*Capricornis crispus*; leaf eating, crop eating), wild boar (*Sus scrofa*; digging up seedlings, crop eating), and Asian black bear (*Ursus thibetanus*; barking, fruit eating, crop eating). The Japanese serow problem was the first problem to occur, and the sika deer problem is one of the most severe. Both problems are good examples with which to examine the relationship between forestry and wildlife because they reflect plantation policy and trends shown by hunters after the Second World War.

Figure 17.1 shows the trends in cutting and plantation from the early Showa era (the Showa era spanned the period 1926-89). The cutting area

Figure 17.1

Trends in cutting and plantation in Japan, 1930-95.

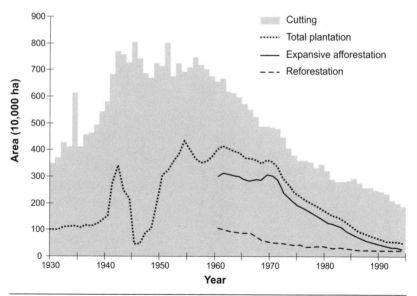

Note: Data before 1960 are not available for expansive afforestation and reforestation.
Source: Forestry Agency, *Statistical Handbook of Forestry* (*Ringyo tokei yoran*), 1930-95.

increased from 1930 to 1940 as a result of the special demand for munitions. From 1940 to 1960, about 700,000 ha were cut every year. In the postwar period, reconstruction of cities required large amounts of wood. Subsequently, the area cut decreased at a constant rate, but it is still nearly 200,000 ha per year. Plantation area became very small at the end of the Second World War, then grew rapidly until 1954. From then to 1970, it was maintained at about 400,000 additional ha every year, but the type of plantation was quite different before and after 1959. Before 1959, expansive afforestation, planting of new areas after natural forests were cut, slightly exceeded reforestation. After 1959, however, expansive afforestation amounted to three to four times as much as reforestation. This radical change was due to the Timber Production Increase Plan. Plantation area has been decreasing since 1970.

Cutting, regardless of plantation, provides more than 10 times more food for herbivorous mammals such as serow and deer as natural forests (Ono et al. 1978; Miura 1992). This condition lasts for several years until undergrowth vegetation becomes poor as the tree crown is closed.

Expansive plantation is commonly recognized as one of the main causes of plantation damage after the Second World War. It provides much better food conditions than natural forest, enabling serow and deer populations to multiply quickly. Cuttings also provided good food conditions. Considering the area of cuttings was much larger than expansive plantation in the 1940s, which was 20 years earlier than the peak of expansive plantation, it is not reasonable to attribute the main cause of damage to expansive plantation. Other conditions must be considered to understand how the problems occurred.

The sizes of serow and deer populations are not clear, and we know only the rough trends. Serow was hunted before the Meiji era. After the Meiji restoration, increased hunting of this species resulted in a marked decrease in the population, which made it necessary to protect these animals (Kaburagi 1938). Serow was dropped from list of game species in 1925; it was designated as a natural monument in 1934 and as a special natural monument in 1955. In the 1920s serow could be found in mountains at high elevations. Their numbers recovered slowly until the 1950s, but they were still rare because of poaching (Chiba 1991). In 1959 the organized poaching of serow was discovered and many hunters were arrested (Ohtaishi 1984). Poaching of this animal subsequently declined, and its numbers have climbed back to the hundreds of thousands. Serow is now one of most common species in mountainous areas (Tokida 1991).

Damage problems related to serow began in the late 1960s. In 1975 the lifting of the ban on capturing serow triggered conflict between protectionists and forestry concerns. The former charged that expansion plantation depleted the serow's habitat, while the latter complained that the

protectionists caused financial loss by interfering with the capturing of
these animals. A management plan for serow known as the Three Authorities Agreements was issued under the joint signatures of the Ministry of
Education, the Environment Agency, and the Forestry Agency. This agreement stated that Serow Preservation Areas were to be established, that serow
were to be designated and protected as natural monuments only in these

Figure 17.2

Trends in issuing hunting licences and killing deer in Japan, 1930-95.

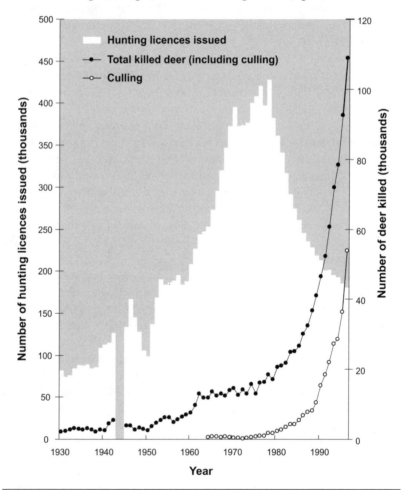

Note: Data for 1944 and 1945 are not available.
Source: Forestry Agency, *Statistical Handbook of Hunting* (*Shuryo tokei*), 1948-62; Forestry Agency,
Statistics of Wildlife Hunted by Hunting Licence Holders (*Suryomenkyosha no chojuhokaku no tokei*),
1923-60; Forestry Agency, *Statistical Yearbook on Hunting and Wildlife* (*Choju Kankei tokei*), 1963-
71; Environment Agency, *Statistical Yearbook on Hunting and Wildlife* (*Choju kankei tokei*), 1972-96.

areas, and that outside of the areas serow should be controlled. After this agreement was published, culling of serow was allowed in two prefectures.

This did not solve the conflict, however, as the forestry concerns were not satisfied because there was no declared date of enforcement. Protectionists, on the other hand, doubted whether suitable and sufficient preservation areas would be established. The situation has become less volatile, although to date no preservation area has been enforced and culling is allowed in five prefectures.

Both female and male deer were hunted until the end of the Second World War. In accordance with GHQ (General Headquarters [of US Army in Japan]) recommendations, female deer were taken off the list of game animals in order to protect the deer population. The number of deer hunted increased rapidly beginning in the 1980s (Figure 17.2). While the number of hunters increased until the 1970s, it has decreased steadily up to the present. Hunting and culling are therefore insufficient to stop the growth of the deer population. Damage to plantations has increased rapidly (Figure 17.3), followed by increases in the numbers of culled animals (Figure 17.2). Culling is an administrative procedure requiring expenditure by the local administrative office. The Environment Agency decided to add female deer to the list of game in 1994, although hunting of female deer is still prohibited in most prefectures.

Figure 17.3

Trends in plantation damage caused by Japanese serow (*Capricornis crispus*) and deer (*Cervus nippon*), 1950-97.

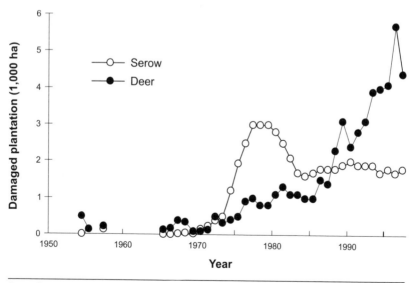

Source: Forestry Agency, *Statistical Handbook of Forestry (Ringyo tokei yoran)*, 1950-97.

Expansion of cutting area in the early Showa era provided good conditions for increasing both serow and deer populations, but these populations remained very low because of exploitative hunting. For several years after the Second World War, it was presumed that the serow population was kept low by poaching. Damage caused by deer peaked later than that of serow in spite of the fact that deer can increase more rapidly than serow from the reproductive point of view, and the protection policy for deer – a ban on hunting female deer – was enforced earlier than that of serow. One cause of this could be the trend in hunters. From the 1960s to 1970s, the number of hunters in Japan doubled. This may have delayed the appearance of the problem. Hunters quickly decreased in number after the late 1970s, and the ban on hunting female deer resulted in an exponential increase in the deer population. In 1994 the ban was lifted in several prefectures. This administrative treatment must be carried out carefully as it may greatly influence the deer population.

Wildlife management and damage problems must be considered from several perspectives: demand for wood, forest policy, trends in hunting, crop protection system, wildlife policy, etc. All these factors, not only population control, must be considered when making wildlife management policy.

Biological Diversity and Conservation of Forests

Social changes related to the public perception of forest resources have resulted in a movement towards wildlife protection. As the forestry industry has declined due to decreased wood prices and increased labour costs, and social demands for environmental conservation have increased, public concern over the roles of forests other than for wood production has increased. These roles are beneficial to the public and add value even to forests with little timber value. Providing wildlife habitats is recognized as one such role of forests, although this recognition brings no real changes in forest operation and/or planning.

A major change occurred after logging problems at Shiretoko National Park in 1986. A controversy arose between the Forestry Agency and environmental conservationists over whether selective cutting should be allowed. Protection of wildlife was one of the biggest issues to arise. Large old trees were scheduled to be cut to accelerate the growth of young trees and regeneration, but environmental conservationists wanted to protect these trees to provide nesting for the endangered large owl, Blakiston's fish-owl (*Ketupa blakistoni*). The trees were eventually cut as planned, but the Forest Agency found it necessary to change the zoning system for forest protection. In 1989 the new protection system was implemented with seven types of protection: forest ecosystem reserves, genetic resource reserves for forest wildlife, genetic resource reserves for tree species, vegetation protection forests, protection forests for specific animal habitats, protection forests for specific landscapes, and hometown forest. The forest ecosystem reserve is influenced by international movements and consists of a core zone and a buffer zone, similar to the biosphere reserve of the UNESCO MAB (Man and Biosphere) project.

Japan has come under a great deal of international pressure to promote species conservation. In October 1991, the director of the EA called for an urgent plan for wildlife protection from the natural environment conservation committee. The report was submitted in February 1992, a month before the international meeting of the Convention on International Trade in Endangered Species of Wild Fauna and Flora (CITES) in Kyoto. This report highlighted the necessity to protect endangered species and promote conservation and countermeasures. It concluded that existing laws and regulations were not enough to protect endangered wild plants and animals. The Biodiversity Convention, which was adopted at the United Nations Conference on Environment and Development in Rio de Janeiro in June 1992, initiated the worldwide trend for species conservation.

These events resulted in the establishment in Japan of the Law for the Conservation of Endangered Species of Wild Fauna and Flora in June 1992. This law aims to protect endangered species and maintain the natural environment in good condition so as to contribute to the quality of human life for both the present and future generations. It designated national rare species and international rare species, and prohibited the capturing, assigning, or trading of these species without permission from the EA. It also designated protected areas to maintain habitats for these species.

Despite this movement towards conservation of wildlife species, the Japanese crested ibis (*Nipponia nippon*), a representative Japanese wildlife species, as the scientific name shows, has become almost extinct. Other species, such as the Japanese otter (*Lutra nippon*), are extremely endangered, indicating that it is very difficult to restore wildlife populations once they become very small.

Movements protecting endangered species have also been directed at other wild animals in Japan. The Asian black bear is a typical case. Trade in the Asian black bear is severely restricted by CITES, which designated it as an Appendix I species. Some local populations in the west of Japan are isolated and small and are in danger of extinction. A few prefectures have prohibited bear hunting despite injurious accidents and damage to crops. There is a "corridor" plan that connects isolated habitats by "green belts" of forests to enable genetic intercourse. Unlike previous schemes for protecting nature by preserving the natural environment, these movements conserve natural resources for the purpose of maintaining the whole environmental system in good functional order.

No convincing or definitive reasons have been put forward for wildlife conservation. We can say only that wildlife is precious or is a part of the ecosystem. Because biodiversity has been recognized as being necessary, however, this concept provides a new way of looking at wildlife conservation (Reid and Miller 1989) – a change from conservation of wildlife to conservation of the whole environment, including mankind.

Wildlife, Forest, and Culture

Although the new keyword *biodiversity* will change wildlife management guidelines, it is still difficult to make social and cultural changes to enable coexistence with wildlife.

For example, even if people agree with conservation of the Japanese serow, it is difficult to determine the optimal population density and who will pay for crop damage or protection costs. From a genetic point of view, a local population must have more than 500 or 1,000 individuals to reduce the possibility of extinction (Washitani and Yawara 1996). These numbers can be used as guidelines only for endangered species. There are no numerical criteria for managing other much more abundant species, which sometimes causes major social problems.

The attitudes or agreements within a local area may provide guidelines for such species. For example, measures to ameliorate the damage caused by the Japanese serow are quite different among different areas. On the one hand, in Gifu and Nagano prefectures, the serow population density is reduced by shooting. On the other hand, in Aomori and Shiga prefectures, crop fields and plantations are protected by fences and serows are not culled. These cases show that the local society has the initiative in choosing a management plan.

To integrate wildlife into local society and culture, an incentive for people to think of wildlife as a resource is necessary. Resource utilization is apt to result in attempts to improve productivity and efficiency, however, leading to an intensive management similar to farming. For coexistence with wildlife in nature, cultural backgrounds are more necessary than resource control.

In the modern world, however, neither city dwellers nor rural residents have frequent opportunities to have direct contact with wildlife. Information about wildlife is obtained mainly from the mass media such as TV and newspapers, and their influences on our attitudes towards wildlife are strong enough to obscure the relationships in the real world. We make humanlike contacts with wildlife through TV programs but control them like inorganic resources in reality.

Other attitudes are appearing slowly, however. A program to reintroduce the white stork (*Ciconia ciconia*), which became extinct in Japan in 1960s due to extermination and habitat deterioration, is currently under way in Hyogo Prefecture. The purpose of this project is to recover the coexistence of farmer and the white storks in the rural environment. There is mounting public opinion that we must positively accept wildlife in our daily life (Sakurai 1995) or re-evaluate the old culture associated with wild animals (Kawai and Hanihara 1995).

Because of the reduced importance of forestry as an industry for wood production, and because of enhanced social consciousness regarding the environment, the functions of forests other than those related to wood production have been highlighted since the 1970s. This trend leads to a comprehensive understanding of activities concerning forests as "forest culture." Furthermore, some say that forests have given rise to growth in our culture. This indicates that forests have more value for modern society than as just a resource. It is the same with wildlife.

It is impossible to accept forests as a resource with no wildlife. It is therefore time for us to create a new wildlife culture. When we can talk about our relationship with wildlife with words other than "management," we will come to know the meaning of coexistence with wildlife.

References
Chiba, Hanji (1991). "Hito to no kakawari." In Omachi Sangaku Hakubutsukan, ed., *Kamoshika*, 119-36. The Shinano Mainichi Shimbun, Nagano (in Japanese).
Endo, Kimio (1994). *Moriokahan on-kari nikki*, 71-77. Kodansha Co., Tokyo (in Japanese).
Environment Agency (1994). "Shuryochoju no shurui wo sadameru ken no ichibu wo kaiseisuru ken tou nitsuite." Official Notification, No. 179 (in Japanese).
Forestry Agency (1969a). *Chojugyosei no ayumi*, 3-6. Rinya Kousaikai, Tokyo (in Japanese).
– (1969b). *Chojugyosei no ayumi*, 6-30, 183-92. Rinya Kousaikai, Tokyo (in Japanese).
Kaburagi, Tokio (1938). "Kamoshika no hozon ni kansuru hiken." In *Tennenkinenbutsu chosa hokokusho*, vol. 2, 85-87. Education Agency, Tokyo (in Japanese).
Kawai, Masao, and Hanihara, Kazurou, eds. (1995). *Doubutsu to bunmei*, 280. Asakura Shoten, Tokyo (in Japanese).
Miura, Shingo (1992). "Shinrin higai wo meguru nihonnkamoshika no 20 nen (I)," *Shinrin Boeki (Forest Pests)* 41(12): 2-8 (in Japanese).
Ohtaishi, Noriyuki (1984). "Kamoshika no kanriho," *Kagaku* 54(1): 50-53 (in Japanese).
Ono, Yuichi; Azuma, Kazutaka; and Doi, Akio (1978). "Sobo-Katamuki sankei ni okeru kamosika no nijirin no riyo nitsuite." In *Tokubetu Tennen Kinenbutsu Kamoshika ni kansuru Chosa-kenkyu-hokokusho (Conservation Report of Japanese Serow, Special Natural Monuments)*, 189-99. Nihon Shizen Hogo Kyokai (Nature Conservation Society of Japan), Tokyo (in Japanese).
Reid, Walter V., and Miller, Kenton R. (1989). *Keeping Options Alive*, 1-8. World Resources Institute, New York.
Sakurai, Masamitsu (1995). "Kinsei shiryo nimiru jugai to sono taisaku no rekishi." In Kawai, M. and Hanihara, K., eds., *Dobutsu to Bunmei*, 52-56. Asakura Shoten, Tokyo (in Japanese).
Takayanagi, Atsushi (1991). "Chojuhogo oyobi shuryo ni kansuru houritsu no kaizen ni tsuite (II)" ("Study on an Improvement of Wildlife Protection and Hunting Law (II)," *Dai 102 Kai Nihon Ringakukai Taikai Ronbunshu (Trans. 102 Mtg. Jap. For. Soc.)*, 313-14 (in Japanese).
– (1993). "Yaseidoubutsu no hogokanri ni kansuru kenkyu," 91-93. Ph.D. dissertation, Department of Agriculture, Kyoto University (in Japanese).
– (1995). "Gendai sanson ni okeru ryoku to shuryokanri." In Kitagawa, M., ed., *Shinrin-*

Ringyo to Chusankanchiiki mondai, 63-89. Nihon Ringyo Chosaki, Tokyo (in Japanese).

Tokida, Kunihiko (1991). "Kamoshika no hogokanri." In Omachi Sangaku Hakubutsukan, ed., *Kamoshika,* 169-78. Shinano Mainichi Shimbun, Nagano (in Japanese).

Tsukamoto, Manabu (1983). "Shorui wo meguru seiji," 8-34. Heibonsha, Tokyo (in Japanese).

Washitani, Izumi, and Yawara, Tetsukazu (1996). *Hozen seitaigaku nyumon (An Introduction to Conservation Biology),* 178-79. Bunichi So-go Shuppan Co., Tokyo (in Japanese).

Contributors

Yoshiya Iwai
Professor, Laboratory of Forest Resources and Society, Division of Forest Science, Graduate School of Agriculture, Kyoto University, Kitashirakawa-oiwake-cho, Sakyo-ku, Kyoto 606-8502, Japan.

Ken-ichi Akao
Associate Professor, School of Social Sciences, Waseda University, 1-6-1 Nishi-Waseda, Shinjuku-ku, Tokyo 169-0051, Japan.

Ichiro Fujikake
Lecturer, Laboratory of Forest Economics, Department of Regional Agriculture Systems, Faculty of Agriculture, Miyazaki University, Gakuen Kibanadai-Nishi 1-1, Miyazaki 889-2192, Japan.

Mitsuo Fujiwara
Professor, Division of Forest Resources, Faculty of Agriculture, Ehime University, Matsuyama 790-8566, Japan.

Kunihiro Hirata
Associate Professor, Laboratory of Forest Policy, Division of Forest Ecology and Management, Department of Environmental Science and Technology, Faculty of Agriculture, Kagoshima University, Koorimoto, Kagoshima 890-0065, Japan.

Takashi Iguchi
Professor, Division of Forestry and Environment, Department of Ecology and Environmental Science, Faculty of Life and Environmental Science, Shimane University, Nishikawatsu-cho 1060 Matsue, Shimane 690-8504, Japan.

Katsuhisa Ito
Professor, Division of Agricultural and Resource Economics, Department of Regional Development, Faculty of Life and Environmental Science, Shimane University, Nishikawatsu-cho 1060 Matsue, Shimane 690-8504, Japan.

Junichi Iwamoto
Assistant Professor, Division of Resource and Environmental Policy, Faculty of Agriculture, Ehime University, Matsuyama 790-8566, Japan.

Koji Matsushita
Associate Professor, Laboratory of Forest Resources and Society, Division of Forest Science, Graduate School of Agriculture, Kyoto University, Kitashirakawa-oiwake-cho, Sakyo-ku, Kyoto 606-8502, Japan.

Shoji Mitsui
Professor, Department of Environmental Science and Technology, Faculty of Bioresources, Mie University, 1515 Kamihama, Tsu 514-8507, Japan.

Hideshi Noda
Group Leader, Forest Management Group, Kansai Research Center, Forestry and Forest Products Research Institute, Momoyama, Fushimi-ku, Kyoto 612-0855, Japan.

Tamutsu Ogi
Professor, Faculty of Management Information, Kyoto-Sosei University, 3370 Hori, Fukuchiyama 620-0886, Japan.

Kozue Taguchi
Assistant Professor, Laboratory of Forest Resources and Society, Division of Forest Science, Graduate School of Agriculture, Kyoto University, Kitashirakawa-oiwake-cho, Sakyo-ku, Kyoto 606-8502, Japan.

Atsushi Takayanagi
Lecturer, Laboratory of Forest Biology, Division of Forest Science, Graduate School of Agriculture, Kyoto University, Kitashirakawa-oiwake-cho, Sakyo-ku, Kyoto 606-8502, Japan.

Kiyoshi Yukutake
Professor, Laboratory of Forest Economics, Department of Regional Agriculture Systems, Faculty of Agriculture, Miyazaki University, Gakuen Kibanadai-Nishi 1-1, Miyazaki 889-2192, Japan.

Index